Magneto-gravitational Coupling and the Missing Mass Problem in Galactic Dynamics

Magneto-gravitational Coupling and the Missing Mass Problem in Galactic Dynamics

By

Babur M. Mirza

Cambridge Scholars Publishing

Magneto-gravitational Coupling and the Missing Mass Problem
in Galactic Dynamics

By Babur M. Mirza

This book first published 2023

Cambridge Scholars Publishing

Lady Stephenson Library, Newcastle upon Tyne, NE6 2PA, UK

British Library Cataloguing in Publication Data
A catalogue record for this book is available from the British Library

ISBN (10): 1-5275-5172-5
ISBN (13): 978-1-5275-5172-5

Contents

Preface

The missing mass problem is now almost a century old and remains one of the major, if not the major, problems of galactic astronomy. The aim of this book is to address the problem of missing mass in galaxies within the context of gravitational coupling to magnetic field in galaxies. I have tried to show that it can be solved using known laws of physics rather than postulating something esoteric and undetected. The process of magneto-gravitational coupling is quite universal and can be shown to describe the large scale structure of the cosmos as well as local effects inside the solar system. Here a systematic attempt is made to show how such a process can be derived from combining Maxwell's electromagnetic theory with the Einstein's general theory of relativity. In simple terms, it is shown that spacetime modifies the magnetic field such that locally magnetic pressure is generated, which in turns acts as an outward drag on test bodies. The magnetic field is wrapped around the gravitational source, and the compressed magnetic field modifies the local energy density of the system. This generates outward moving spiral (magneto-gravitational) waves characteristic of the spiral galaxies. In the initial stages of galaxy formation, these structures are weak due to the weak gravitation and magnetic field.

This book consists of five chapters. The first chapter gives some basic information about galaxies and their astronomical and astrophysical characteristics. In the same chapter an over-view of relativistic frame dragging (Lense-Thirring dragging) is given. Chapter 2 gives a brief review of the general theory of relativity, in particular the Einstein field equations and introduces the spacetime metric for rotating gravitational masses. Chapter 3 gives a mathematical background of the generalized Maxwell equations in the spacetime of rotating massive source. The electromagnetic field is determined for the case when the rotating mass has rotational speed far below the relativistic speed limit. This gives the dominant effect of the gravitational field in the magneto-gravitational coupling. Chapter 4 applies the magneto-gravitational coupling for the toroidal magnetic field to the case of spiral galaxies. The typical size and age, as well as the rotation curves of different galaxies, are obtained from the model. This is done without invoking a density function corresponding to the dark matter component. The galactic mass is found to be consistent with the mass needed to generate the kinetic effects otherwise corrected by the missing galactic mass employing the mass distribution function. Chapter 5 gives a summary of the main conclusions of the work. Throughout I have tried to maintain continuity of presentation, but have presumed that the reader is acquainted with both General Relativity and Classical Electromagnetic Theory in tensor form.

September 2023 *Babur M. Mirza*

Chapter 1

Introduction

1.1 Galaxies - Astrophysical Features

Galaxies are large astrophysical systems of stars bound together by gravitation. Unlike stars, which are spherical or depart very little from spherical shape, galaxies exist in many shapes from spherical ones to highly flattened. On average a galaxy contains about $10^{10} - 10^{12}$ stars. The Milky Way (Galaxy) contains about $1-4 \times 10^{11}$ stars. The galactic age falls within the range of

$1 - 2 \times 10^{10} yr$. Galaxies are formed by condensation of intergalactic gas, and are more luminous in their early formation period. Galactic size (radius) is usually measured in kpc, where $1kpc = 3.09 \times 10^{16} m$. The Milky Way is approximately $26.8kpc$ in radius. There are two common catalogues for galaxies. The Messier catalogue refers to galaxies with Messier numbers such as M105 or M99. The other catalogue commonly used is the New General Catalogue of Nebulae and Clusters of Stars (NGC), it numbers galaxies as NGC 3980, etc.

In 1926 E. Hubble (Hubble 1926) gave a classification of galaxies based on their overall appearance. In Hubble's morphological classification scheme (also called Hubble's sequence), galaxies are grouped into three primary categories: ellipticals (E's), spirals, and irregulars (Irr's). The spiral galaxies are further subdivided into normal spirals (S's), and barred spirals (SB's). A further class of galaxies, known as lenticulars, is a transitional class of galaxies between ellipticals and spirals. The lenticular galaxies are either normal (S0's) or barred (SB0's). Hubble morphological sequence is often represented as a tuning-fork diagram (Figure 1.1), where the two types of spiral galaxies can be clearly indicated.

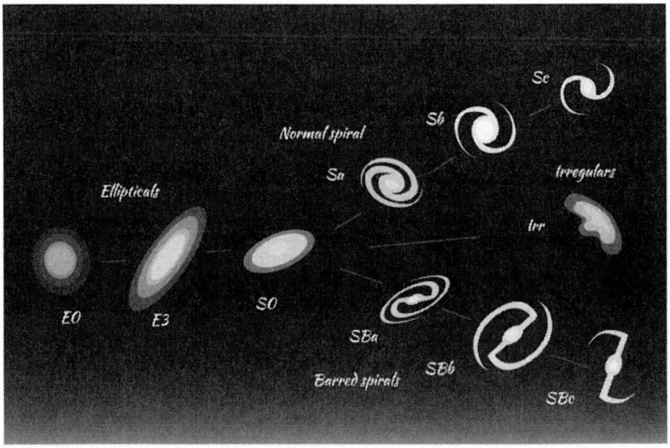

Figure 1.1: Hubble's tuning-fork diagram classifies galaxies into main three types: ellipticals (E's), spirals, and irregulars (Irr's). (Image is from www.vectorstock.com)

The average diameter of elliptical galaxies lies between $0.1 - 10^2 kpc$. The giant elliptical galaxies are among the largest known objects in the universe, where as the dwarfs ellipticals have a size comparable to a typical globular cluster. Elliptical galaxies have the absolute magnitude B as low as 8 and as high as 23. Their masses too cover a large range from $10^7 M_\odot$ to$10^{13} M_\odot$. The lenticular galaxies have masses and luminosities comparable to the large elliptical galaxies. This makes the lenticular galaxies, along with the large elliptical galaxies, among the easiest to observe, however, the dwarfs are by far the most common type of galaxies.

Hubble's classification further subdivides the spiral galaxies into the sequence Sa, Sab, Sb, Sbc, Sc, and SBa, SBab, SBb, SBbc, and SBc. The Sa (or SBa) type of spiral galaxies have the most prominent bulges as compared to the disk size. They have the most tightly wound spiral arms. The Sc (or SBc) type galaxies have smaller bulge-to-disk ratios and more loosely wound spiral arms. $M31$ and $NGC891$ galaxies are Sb's, whereas the Milky Way is probably an SBbc. The spiral galaxies tend to have much smaller variations in their physical parameters (mass $10^9 M_\odot - 10^{12} M_\odot$, radius $5 - 10^2 kpc$, absolute magnitude $B = 16 - 23$) than do ellipticals. Spiral galaxies are also among the largest galaxics in the universe.

The irregulars galaxies are divided into Irr I with some organized structure, and Irr II with the completely disorganized structures. The Large Magellanic Cloud and the Small Magellanic Cloud are Irr I galaxies, where as M82 is an example of an Irr II. Irregular galaxies are not very large. Their B magnitude vary from -13 to -20, they have masses of between $10^8 M_\odot$ and $10^{10} M_\odot$, and their diameters range from 1 to $10kpc$. Irregular galaxies often have noticeable bars also.

1.2 Relativity and Spacetime

In Newtonian mechanics absolute motion is a motion from one absolute place to another. The absolute places taken together constitute a space that is fixed through time. The motion of bodies with respect to each other is relative motion, where as motion with respect to the absolute space is absolute motion. The material example of an absolute space is the frame of reference defined by the distant 'fixed stars'. However the distant stars may be in constant motion (uniform linear velocity) or at rest, because in Newtonian mechanics both states are the same and define an inertial frame. In classical (Newtonian) mechanics, therefore, distant stars do not provide a reliable means of distinguishing between absolute and relative motion. Then does absolute motion has any meaning at all?

Newton reversed the argument and asserted that inertial forces define an inertial frame (that is, absolute motion). Now inertial frame is taken as the frame of reference in which Newton's laws (particularly the second law of motion $F = ma$) hold. To prove that inertial forces exist, thus inertial frames (hence absolution motion) exist, Newton devised the classic 'bucket' experiment. In this experiment a bucket containing water is suspended with its axis of symmetry vertical. When not is motion about its axis (rotation), the water surface is flat relative to the observer (defining a laboratory frame or lab frame). But when set in rotation about the vertical axis, the water surface becomes concave with the rising speed of the bucket. Newton saw this as a verification of absolute space, because the inertial force (centrifugal force in this case) existed only when the bucket was rotated and subsided when the bucket came to rest. Even when the bucket was held fixed at later stages of the experiment the water continued to creep up the walls of the bucket. Thus the validity of inertial forces was established.

In the late nineteenth century, Ernst Mach criticized Newton's reasoning on the following grounds. Mach contended that absolute space and absolute motion are mental constructs and have no correspondence with actual experience. Mach argued that since in the rotating bucket experiment the observer is not rotating with the water, the centrifugal force appears. For an observer defining a correct frame (rotating with water locally), there will be no way to distinguish locally (that is without referring to another frame) the motion of water. The water surface curves only in the relative frame of the outside observer. Mach gave another specific case of the rotation of the Earth relative to the distant stars. In Newtonian mechanics it is immaterial whether the Earth spins around its own axis or the fixed stars revolve around a fixed Earth with the same angular speed. But phenomena like the oblate form of the Earth, and the rotation of the plane of the Foucault's pendulum, occur only when one assumes (as in Newtonian mechanics) that the Earth rotates and not the distant stars. Mach concluded that only relative spaces and motion (as well as masses) are possible (as given facts). Mach held that, "When accordingly, we say that a body preserves unchanged its direction and velocity *in space*, our assertion is nothing more or less than an abbreviated reference to *the entire universe.*"(Mach 1893). Mach's conjecture was taken up by Einstein to imply that inertia like other forces in Newtonian mechanics must depend on the interaction between masses. Mach, however, has also considered the possibility of inertial forces arising *not* from direct interaction of masses, but from the interaction of mass with a medium. This medium, Mach asserted, could be a substitute for Newton's absolute space.

The problem of non-existence of an absolute frame of reference (Ether) led Einstein to built his special theory of relativity (Einstein 1952) based on the equality of all inertial frames of reference, without any privileged fixed (universal) frame, such as held in Lorentz's electron theory. Assuming that the speed of light is an absolute speed, Einstein showed that the paradoxes appearing in Maxwell-Lorentz electrodynamics, such as the asymmetry arising in interpreting Faraday induction experiments, do not occur. It was also shown that Maxwell's equations for the electromagnetic field preserve their form (covariance) in the new theory of motion, provided that time is also taken as a relative rather than as an absolute measure of change. This similarity of space and time measures and their relativity led Herman Minkowski (see Einstein 1952) to construct a four-dimensional geometry of space-time, now known as Minkowski spacetime geometry. In this geometry particle trajectories (geodesics) are straight lines lying within the light cone, with different slopes. These represent constant uniform motions of special relativity. Later, to incorporate gravitation, Einstein saw that accelerated motions have to be introduced in this geometry[1]. He thus introduced the notion of curved spacetime, caused by matter sources, in which geodesics give the particle trajectories. In this way

[1]The impetus came from Galileo's free-fall experiment. The equivalence of inertial mass (resistance to motion) and gravitational mass (determined by motion due to gravity) gave Einstein the idea of motion in a gravitational field, whose effects in the natural state of motion (geodesic) vanish. This is analogous to the case of electromagnetic field, particularly the induction experiments.

Mach's medium was replaced with spacetime continuum through which material objects interact gravitationally. Based on this Einstein derived the perihelion shift of Mercury in 1916, and predicted the deflection of light by the Sun. This prediction was verified by A. Eddington and his team in 1919 (Eddington 1919).

1.3 The Frame-Dragging Effect

In the Einstein's theory of relativity, frames of reference locally define the properties of space. Thus Newtonian mechanics is locally valid because space (spacetime) is locally flat, except in extremely high energy concentration regions such as those in vicinity of compact gravitational sources. In weakly gravitating systems or in the absence of a gravitational source, the spacetime is the Minkowski spacetime. However, if the motion is slow compared with the speed of light, the spacetime coupling is weak, and spacetime separates into the ordinary space (of three dimensions) along with an independent time component, as in classical Newtonian mechanics.

If the analogy of spacetime or gravitational field

with electromagnetic field is pursued further, the effects of a gravitational source can be identified by the local behavior of various frames of reference at each point in the spacetime around the source mass. This analogy was first used by Einstein to deduce (Einstein 1912) what later came to be called the frame-dragging effect. Einstein referred to the induced motion of a charge in vicinity of a current carrying conductor, and concluded that a test mass in vicinity of a rotating mass would too be set in motion due to the dragging of the inertial (local) frames. Later on, when full equations of gravitational field were obtained by Einstein, Lense and Thirring (Lense & Thirring 1918) showed that such an effect can be derived from the linearized Einstein field equations also. After many attempts to solve Einstein field equations for a rotating massive object, R. Kerr (Kerr 1963) finally obtain the full solution (spacetime metric) for this case.

Considering the fixed stars as asymptotically flat spacetime reference frame, the change in the plane of oscillations of a Foucault pendulum at the Earth's North pole, is due to the locally produced rotation of the Earth. However, such a change will not be detectable in a co-moving frame. Only a relative motion between the two, such as between the Earth and that referred to the distant stars, can discern the effect. Analogously, the spacetime in general is dragged around a rotating mass for an asymptotically distant observer not rotating with the source. This effect, the Lense-Thirring frame dragging of space (inertial frames), has negligible magnitude around weakly gravitating systems such as the Earth and the Sun. However, attempts are being made to detect it in high precision experiments (see, e.g., Iorio 2010) inside the solar system.

1.4 The Missing Mass Problem

Although stars in a galaxy are constantly in motion, the average number of stars, hence its total mass, is relatively fixed. Therefore overall mass distribution in a galaxy is constant in time (hence gravitational force is time-independent). However in most galaxies the mass distribution is spatially extended, and can be generally divided into three regions: the galactic disk, which contain stars, gas, and dust that is mostly confined to a thin, rotating disk; and the bulge, a central mass accumulation where mass density is higher. Spiral arms reside in the galactic disk, and the bar like structure belongs to the inner galactic bulge region. In the galactic disk, the stars move in nearly circular (coplanar) orbits. The disk therefore rotates collectively. A rotation curve is a plot between the orbital speed of stars and their position in the galaxy.

Since galactic systems are bound by the gravitational force, the Kepler's law of orbital motion holds, hence the galactic rotation curves must show a decrease in the outer part of galactic disks. However, with further improvements in observational astronomy, it was found in the 1970s (see, Rubin 1983, also, Sofue & Rubin 2001) rotation curves show large deviations from the predicted values. Some galactic rotation curves fall-off differently, others remain flat, and still others show a rise at large radii. The most recurrent are the flat or nearly flat rotation curves. The higher observed rotation speed implies that the gravitational force is stronger at large galactic radii, which indicates that there must be more mass than expected. Evidence for the "missing mass" appeared as early as the 1930s, from an analysis of motions in the Coma cluster of galaxies by F. Zwicky (Zwicky 1937). Also,

J. Oort (Oort 1932) deduced the mass difference in an analysis of vertical motions of stars in the Milky Way. However, definite conclusions could not be drawn due to the imprecise measurement of the mass-to-light ratios at that time. Since the "missing mass" component could not be detected in the luminosity measurements, it was considered a non-emitting form of matter, latter named as the dark matter component. This missing mass was found to contain approximately six times the observed mass in a galaxy.

One resolution of the missing mass problem is to assume galactic disk to be embedded in a dark matter halo. The dark matter halo can be modelled, based on the observed rotation curve of a galaxy, as spherical, isothermal, etc. The assumed density profiles then are used to fit the mass function $M(r)$, hence to correct the Newtonian mass function for the velocity curve. Another alternative had been to assume that the Newton's law of forces becomes invalid at galactic scales. The modified Newtonian dynamics (MOND) assumes either a change in the classical Poisson equation, or a breaking of the equivalence of inertial and gravitational mass. Still other alternatives include: Cold Dark Matter, WIMPs (Weakly Interacting Massive Particles), Axions; massless neutrinos, WDM (Warm Dark Matter), MACHOs (Massive Astronomical Compact Halo Objects), dead stars including dwarfs, neutron stars, black holes. Further theoretical models to account for the missing galactic mass can be divided into alternative gravitational theories, or models based on plasma dynamics. These include quantum gravity and bubble force, conformal Weyl gravity, plasma cosmology, etc. However, these models either contradict or do not account for other observed features of galaxies, such as the Tully-Fisher relation, the galactic size and age.

Chapter 2

Gravitational Field

2.1 Metric and Gravitational Field

Minkowski showed that the basic difference between Newtonian mechanics and Special Relativity is due to the metric used for the geometry of space. That is, where as the geometry of the former is Euclidean, for the latter it must be that of a four-dimensional spacetime. This is expressed by defining the metric for special relativistic spacetime. Minkowski showed that if one were to assume that the spacetime metric has the form:

$$ds^2 = \eta_{\alpha\beta} dx^\alpha dx^\beta, \tag{2.1}$$

where $\alpha, \beta = 1, 2, 3, 4$, such that $(x^\alpha) = (ct, x, y, z)$ and

$$\eta_{\alpha\beta} = \begin{bmatrix} +1 & 0 & 0 & 0 \\ 0 & -1 & 0 & 0 \\ 0 & 0 & -1 & 0 \\ 0 & 0 & 0 & -1 \end{bmatrix} \equiv diag\,[+1, -1, -1, -1], \tag{2.2}$$

then the Lorentz transformations, hence all the consequence of Einstein's special theory of relativity, including length contraction and time dilation effects, could be deduced just from the covariant character of equation (2.1). By a *covariant form* or equation we mean an equation or a relation that takes the same form under a coordinate transformation. Sometimes the word invariant is used. For example, a tensor equation is a covariant equation under arbitrary change of coordinates.

In the general theory of relativity, gravitational force is regarded as a manifestation of the curvature of spacetime. Thus any process occurring in vicinity of a gravitating source is affected by the geometry of the background spacetime. Basic to the geometrical structure of the theory is the general line element given by

$$ds^2 = g_{\alpha\beta} dx^\alpha dx^\beta, \tag{2.3}$$

where $g_{\alpha\beta}$ is the metric tensor defined in terms of products of the partial derivatives of the coordinate transformation. Physically the components of the metric tensor represent the gravitational field potentials in a four dimensional spacetime. From these components the geometry of the spacetime is determined by the distribution of the matter content acting as the source of the gravitational field. This is similar to the electromagnetic field theory where the electromagnetic field is determined by the distribution of charges. Of particular importance for spacetime curvature is the Riemann-Christoffel curvature tensor defined by

$$R^\alpha_{\beta\gamma\delta} = \Gamma^\alpha_{\beta\delta,\gamma} - \Gamma^\alpha_{\beta\gamma,\delta} + \Gamma^\eta_{\beta\delta}\Gamma^\alpha_{\gamma\eta} - \Gamma^\eta_{\beta\gamma}\Gamma^\alpha_{\delta\eta}, \tag{2.4}$$

where the Christoffel symbol $\Gamma^\alpha_{\beta\gamma}$ is related to the metric tensor by

$$\Gamma^\alpha_{\beta\gamma} = \frac{1}{2} g^{\alpha\lambda} \{g_{\beta\lambda,\gamma} + g_{\lambda\gamma,\beta} - g_{\gamma\beta,\lambda}\}. \tag{2.5}$$

The fundamental equations of the general theory of relativity can now be expressed as

$$R_{\alpha\beta} - \frac{1}{2} g_{\alpha\beta} R = \kappa T_{\alpha\beta}, \tag{2.6}$$

where $R_{\alpha\beta} = R^{\lambda}_{\alpha\lambda\beta}$ and $R = g^{\alpha\beta} R_{\alpha\beta}$ is the Ricci tensor and Ricci scalar obtained by contraction of the Riemann-Christoffel tensor and κ is a constant of proportionality equal to $8\pi G/c^4$ in S.I. units; whereas the matter content is described by the energy-momentum tensor for a perfect fluid by

$$T_{\alpha\beta} = (p + \epsilon) u_\alpha u_\beta + p g_{\alpha\beta}, \tag{2.7}$$

where p is the pressure, $\epsilon = \rho c^2$(equal to ρ in gravitational units) is the proper energy density, and $u_\alpha = dx_\alpha/ds$ is the velocity four-vector. Hence the left hand side of the gravitational field equation expresses the effect of spacetime curvature via the Riemannian curvature tensor and the right hand side represents the effects of matter content. It is clear, due to the symmetry of $T_{\alpha\beta}$ in the indices α and β, that these are ten coupled highly non-linear partial differential equations.

The motivation for these equation comes from the fact that Special Relativity is confined to non-accelerated motions. For more general types of motion, particularly rotation, Einstein postulated that the acceleration of the reference system is equivalent to a gravitational field. Thus, in his search for a theory of gravitational field, Einstein was led by the following demands:

(1) In gravitation-free case, the field equations must be invariant under the Lorentz transformation. In other words, these equations must be invariant at least under a linear transformation of coordinates.

(2) The curvature of space (spacetime) must relate to the gravitational field. In other words the presence of mass and its distribution must determine this curvature.

(3) These equations must reduce to the Newton's law of gravitation under appropriate limiting conditions.

In analogy with the charge continuity equation, one should assume a current 4-vector for gravitational field as well. But since the spacetime is determined by a second rank tensor $g_{\alpha\beta}$, the tensor describing matter distribution and causing the space to curve accordingly must be of the same rank. Einstein was thus led to the following formulation: the gravitational field equations should be of the form

$$\mathcal{E}_{\mu\nu} = \kappa T_{\mu\nu}, \tag{2.8}$$

where κ is a constant, and $\mathcal{E}_{\mu\nu}$ is a tensor of the second rank, which is generated by differential operations from the fundamental tensor $g_{\mu\nu}$. We note that in analogy with charge conservation equation, matter-energy conservation must result from the condition that the tensor $T_{\mu\nu}$ representing the matter part must have zero divergence. In curved spacetime, this requirement is expressed as

$$T_{\mu\nu;\nu} = 0. \tag{2.9}$$

In his initial work (1912 – 1913) in collaboration with Marcel Grossmann (Einstein & Grossmann 1913), Einstein started from the third condition. He noted that the Newton's law of gravitation (in the form of Poisson equation)

$$\nabla^2\Phi = 4\pi G\rho, \tag{2.10}$$

where $\Phi(\mathbf{x})$ is the gravitational potential and $\rho(\mathbf{x})$ is the mass density, contains the second partial derivatives with respect to the coordinates. An analogous condition for the derivatives of metric tensor therefore results.

2.2 Einstein Field Equations

The program of the general theory of relativity is to obtain equations by which the gravitational field of a given massive object can be determined. These can be obtained using the principle of least action also. First we determine the action function S_g for the gravitational field. Analogous to the electromagnetic theory where the field equations are obtained by requiring the action to be expressed by a *scalar* integral taken over all space and over the time coordinate which must be invariant under a coordinate transformation; further the requirement that the gravitational field equations must not contain derivatives of the potentials (i.e. $g_{\mu\nu}$) higher than the second. This in turn implies that the integrand of the action integral must not contain derivatives of $g_{\mu\nu}$ higher than the first order. With these requirements the suitable choice of the action function S_g is given by

$$S_g = \int R\sqrt{-g}d\Omega, \tag{2.11}$$

where $\sqrt{-g}d\Omega$ the four dimensional volume element and R is the Ricci scalar given by

$$R \equiv g^{\mu\nu}R_{\mu\nu} = g^{\mu\nu}(\partial_\lambda \Gamma^\lambda_{\mu\nu} - \partial_\nu \Gamma^\lambda_{\mu\lambda} + \Gamma^\lambda_{\mu\nu}\Gamma^\eta_{\lambda\eta} - \Gamma^\eta_{\mu\lambda}\Gamma^\lambda_{\nu\eta}). \tag{2.12}$$

Here $\Gamma^\lambda_{\mu\nu}$ is a pseudo tensor formed out of the metric tensor and its first derivatives:

$$\Gamma^\lambda_{\mu\nu} \equiv \frac{1}{2}g^{\lambda\eta}(\partial_\nu g_{\eta\mu} + \partial_\mu g_{\eta\nu} - \partial_\eta g_{\mu\nu}). \tag{2.13}$$

Although the action function S_g contains the second derivatives of the metric tensor, as we shall see in the following the field equations turn out to contain only the first partial derivatives of $g_{\mu\nu}$.

The next step now is to determine the action function S_m due to the matter content. Let us postulate its form as

$$S_m = \int \Lambda\sqrt{-g}d\Omega, \tag{2.14}$$

where Λ is an arbitrary scalar function. The gravitational field equations are now obtained by the least action principle

$$\delta(S_g + S_m) = 0. \tag{2.15}$$

Let us first apply the variation to the gravitational field, that is, to the quantities $g_{\mu\nu}$, we have

$$\begin{aligned} \delta S_g &= \delta \int R\sqrt{-g}d\Omega \\ &= \int (R_{\mu\nu}\sqrt{-g}\delta g^{\mu\nu} + R_{\mu\nu}g^{\mu\nu}\delta\sqrt{-g} + \sqrt{-g}g^{\mu\nu}\delta R_{\mu\nu})d\Omega \end{aligned} \tag{2.16}$$

Using the fact that $\delta\sqrt{-g} = -\frac{1}{2}\sqrt{-g}g_{\mu\nu}\delta g^{\mu\nu}$, we obtain after some simplification

$$\delta S_g = \int (R_{\mu\nu} - \frac{1}{2}g_{\mu\nu}R)\delta g^{\mu\nu}\sqrt{-g}d\Omega + \int g^{\mu\nu}\delta R_{\mu\nu}\sqrt{-g}d\Omega. \tag{2.17}$$

Here, applying the Gauss' theorem, the second term can be written as an integral over a hypersurface enclosing the four volume

$$\int g^{\mu\nu}\delta R_{\mu\nu}\sqrt{-g}d\Omega = \int \partial_\lambda(\sqrt{-g}w^\lambda)d\Omega, \qquad (2.18)$$

where the four vector w^λ is defined as $w^\lambda \equiv g^{\mu\nu}\Gamma^\lambda_{\mu\nu} - g^{\mu\lambda}\Gamma^\nu_{\mu\nu}$, such that $g^{\mu\nu}\delta R_{\mu\nu} = \partial_\lambda w^\lambda$. Since the variations of the field are zero at the integration limits, this integral vanishes identically and in consequence the second term in equation (2.17) drops out.

Now take the variation of the action function S_m. We have

$$\begin{aligned}\delta S_m &= \delta\int \Lambda\sqrt{-g}d\Omega = \int\{\frac{\partial(\sqrt{-g}\Lambda)}{\partial g^{\mu\nu}}\delta g^{\mu\nu} + \\ &\quad \frac{\partial(\sqrt{-g}\Lambda)}{\partial(\partial_\lambda g^{\mu\nu})}\delta(\partial_\lambda g^{\mu\nu})\}d\Omega. \qquad (2.19)\end{aligned}$$

Here the second term in the integrand is expressible as

$$-\partial_\lambda\{\partial(\sqrt{-g}\Lambda)/\partial(\partial_\lambda g^{\mu\nu})\}\delta g^{\mu\nu},$$

where there is a summation over the repeated index λ. We introduce the notation

$$T_{\mu\nu} \equiv \frac{2}{\sqrt{-g}}\{\frac{\partial(\sqrt{-g}\Lambda)}{\partial g^{\mu\nu}}\delta g^{\mu\nu} - \partial_\lambda(\frac{\partial(\sqrt{-g}\Lambda)}{\partial(\partial_\lambda g^{\mu\nu})})\}, \quad (2.20)$$

where $T_{\mu\nu}$ is the energy-momentum tensor to within a constant factor. The variation δS_m becomes

$$\delta S_m = \int T_{\mu\nu}\delta g^{\mu\nu}\sqrt{-g}d\Omega. \quad (2.21)$$

Combining these results we obtain

$$\int (R_{\mu\nu} - \frac{1}{2}g_{\mu\nu}R - \kappa T_{\mu\nu})\delta g^{\mu\nu}\sqrt{-g}d\Omega = 0, \quad (2.22)$$

where κ is a constant of proportionality given by $\kappa = 8\pi G/c^4$ in S.I. units. Since $\delta g^{\mu\nu}$ is an arbitrary variation, we obtain the gravitational field equations:

$$R_{\mu\nu} - \frac{1}{2}g_{\mu\nu}R = \kappa T_{\mu\nu}. \quad (2.23)$$

It is clear, due to the symmetry of $T_{\mu\nu}$ in the indices μ and ν, that these are ten non-linear partial differential equations.

For a weakly gravitating objects the set of equation (2.23) can be reduced to an approximate form. This approximation is called *linearization of the field equations*. Here the metric is assumed to be as for Minkowski spacetime with perturbations introduced by the presence of a rotating massive gravitational source. Thus the metric tensor is expressed as

$$g_{\alpha\beta} = \eta_{\alpha\beta} + h_{\alpha\beta}, \tag{2.24}$$

where $\eta_{\alpha\beta}$ is the Minkowski metric tensor of signature $(+,-,-,-)$ with the inverse $\eta^{\alpha\beta} = \eta_{\alpha\beta}$, and $h_{\alpha\beta}$ is the perturbation to the metric. Under the linear approximation we regard $h_{\alpha\beta} \ll 1$, so that the product of the $h_{\alpha\beta}$ can be neglected as small. Further it is imposed that the second derivatives of $h_{\alpha\beta}$ are also small and negligible. Physically this means that the effects of spacetime curvature (the tidal effects), represented by the second order derivatives of the metric tensor, are negligible. With these approximations, the Christoffel symbols (2.5) can be written as

$$\Gamma^{\alpha}_{\beta\gamma} = (h^{\alpha}_{\beta,\gamma} + h^{\alpha}_{\gamma,\beta} - h_{\beta\gamma,}{}^{\alpha})/2. \tag{2.25}$$

For the Ricci tensor we neglect the products of the

Christoffel symbols since these include the products of $h_{\alpha\beta}$. We then obtain

$$R_{\alpha\beta} = (\Box h_{\alpha\beta} + h_{,\alpha\beta} - h^{\gamma}_{\alpha,\beta\gamma} - h^{\gamma}_{\beta,\alpha\gamma})/2, \tag{2.26}$$

where

$$\Box \equiv_{,\alpha\beta} \eta^{\alpha\beta} \equiv \frac{\partial^2}{\partial t^2} - \frac{\partial^2}{\partial x^2} - \frac{\partial^2}{\partial y^2} - \frac{\partial^2}{\partial z^2}. \tag{2.27}$$

Let us now introduce the auxiliary variables $\zeta_{\alpha\beta}$ such that

$$\zeta_{\alpha\beta} \equiv h_{\alpha\beta} - \frac{1}{2}\eta_{\alpha\beta}h, \tag{2.28}$$

where

$$h = \eta_{\alpha\beta}h^{\alpha\beta}. \tag{2.29}$$

In this way the term $h_{,\alpha\beta}$ disappears and we have for the Ricci tensor $R_{\alpha\beta}$ the following form

$$R_{\alpha\beta} = (\Box h_{\alpha\beta} - \zeta^{\gamma}_{\alpha,\beta\gamma} - \zeta^{\gamma}_{\beta,\alpha\gamma})/2, \tag{2.30}$$

We can further put (2.30) into a more manageable form by transforming the coordinates x^α into $\tilde{x}^\alpha$ as follows

$$\tilde{x}^\alpha = x^\alpha + \lambda^\alpha(x), \tag{2.31}$$

where λ^α and its derivatives have the same order as $h_{\alpha\beta}$. The metric tensor transforms according to the rule

$$g_{\alpha\beta} = \tilde{g}_{\rho\sigma}\frac{\partial\tilde{x}^\rho}{\partial x^\alpha}\frac{\partial\tilde{x}^\sigma}{\partial x^\beta} = \tilde{g}_{\rho\sigma}(\delta^\rho_\alpha + \lambda^\rho{}_{,\alpha})(\delta^\sigma_\beta + \lambda^\sigma{}_{,\beta}). \tag{2.32}$$

The $h_{\alpha\beta}$ then transform as

$$\tilde{h}_{\alpha\beta} = h_{\alpha\beta} - \lambda_{\alpha,\beta} - \lambda_{\beta,\alpha}. \tag{2.33}$$

It therefore follows that

$$\Box\zeta^\alpha = \lambda^{\alpha\beta}{}_{,\beta}\,. \tag{2.34}$$

Applying the Lorentz gauge condition

$$\zeta^{\alpha\beta}{}_{,\beta} = 0, \tag{2.35}$$

we obtain from equation (2.35)

$$\Box\zeta^\alpha = 0. \tag{2.36}$$

Thus in the Ricci tensor the last two terms vanish and we have $2\tilde{R}_{\alpha\beta} = \tilde{\Box}\tilde{h}_{\alpha\beta}$. Now the tilde can be omitted, and hence we obtain

$$R_{\alpha\beta} = \frac{1}{2}\Box h_{\alpha\beta}. \tag{2.37}$$

Therefore the Ricci scalar is

$$R = \eta^{\alpha\beta}R_{\alpha\beta} = \frac{1}{2}\Box h. \tag{2.38}$$

From the last two results the left hand side of the gravitational field equation can be expressed as

$$\Box(R_{\alpha\beta} - \frac{1}{2}\eta_{\alpha\beta}R) = \frac{1}{2}\Box(h_{\alpha\beta} - \frac{1}{2}\eta_{\alpha\beta}h) = \frac{1}{2}\Box\zeta_{\alpha\beta}. \tag{2.39}$$

We thus obtain for the gravitational field equations in the linearized form

$$\Box\zeta_{\alpha\beta} = -2\kappa T_{\alpha\beta}, \tag{2.40}$$

or equivalently

$$\Box h_{\alpha\beta} = -2\kappa(T_{\alpha\beta} - \frac{1}{2}\eta_{\alpha\beta}T). \tag{2.41}$$

This shows that $h_{\alpha\beta}$ and $T_{\alpha\beta}$ are of the same order of smallness, so that the products of $h_{\alpha\beta}$ and $T_{\alpha\beta}$ can be ignored. Equations (2.40) and its equivalent form (2.41) are linear field equations, hence the name *linear approximation* for this theory.

In the *Newtonian limit*, that is, in a reference frame moving with velocity V that is small as compared to the velocity of light, the rest energy dominates the stress-energy of the non-relativistic matter. In such a case the spacetime curvature produced by the rest energy density ρ is small. Consequently the kinetic energy ($\rho V^2/c^2$ in S.I. units) and the potential energy (i.e. $\rho\Phi/c$) are small. Thus $T^{\alpha\beta}$ with $u^\alpha \approx (-1, \mathbf{0})$ reduces to

$$T_{\mu\nu} = \begin{bmatrix} \rho & 0 & 0 & 0 \\ 0 & 0 & 0 & 0 \\ 0 & 0 & 0 & 0 \\ 0 & 0 & 0 & 0 \end{bmatrix}, \tag{2.42}$$

and therefore we have

$$T_{\mu\nu} - \frac{1}{2}\eta_{\mu\nu}T = \frac{\rho}{2}\begin{bmatrix} 1 & 0 & 0 & 0 \\ 0 & 1 & 0 & 0 \\ 0 & 0 & 1 & 0 \\ 0 & 0 & 0 & 1 \end{bmatrix}, \tag{2.43}$$

we obtain from (2.40) that

$$h_{00} = h_{11} = h_{22} = h_{33}. \tag{2.44}$$

Since the motion of sources is slow, the gravitational field changes slowly also, so that $\Box h_{\alpha\beta} \approx -\nabla^2 h_{\alpha\beta}$. This gives in the linearized field (2.41)

$$\nabla^2 h_{00} = -\kappa\rho. \tag{2.45}$$

The solution to the last equation is the $h_{00} = 4\Phi$, where Φ is given by

$$\Phi = -\int\int\int \frac{\rho}{r} dV. \tag{2.46}$$

For a mass point source of mass M, (2.46) is the usual Newtonian potential

$$\Phi = -\frac{M}{r}. \tag{2.47}$$

Hence the Newtonian gravitational law is a special case of the gravitational field equations if the sources move negligibly slowly, and have negligible stress-energy.

In ordinary circumstance, such as terrestrial conditions and inside the solar system, gravitational effects can be adequately described to a very high degree of accuracy within the Newtonian theory of gravitation. Thus it is assumed that the gravitational force depends on the mass of the interacting bodies and their mutual distances only. For strongly gravitating objects, such as compact objects, it is necessary, to include general relativistic effects. In fact the very possibility of theses stars is due to general relativistic effects. However in the general theory of relativity, where spacetime curvature plays a fundamental role, the nature of these effects can be very different from the Newtonian theory. For instance the gravitational force, contrary to the Newtonian theory, depends on the velocity of the rotating gravitational source (Lense-Thirring effect), on the spin of a relativistic test particle (De Sitter precession), on the generation of gravitational waves, etc. Interestingly, though these effects have no analogues in the Newtonian theory, these effect have equivalents in the classical electromagnetic theory (induced magnetism, Thomas precession, electromagnetic waves). This analogy has been found to be correct provided that the field of the interacting masses is not extremely intense (such as for totally collapsed objects) and the gravitational source is rotating slowly (that is within the special relativistic limits on speeds). The study of these effects in compact stars and their vicinities is of particular interest also for the reason that the detection of theses effects provides a means for testing General Relativity. On Earth and inside the solar system gravitomagnetic effects are extremely small, thus requiring extremely precise measurements.

2.3 Non-rotating and Rotating Masses

Schwarzschild assumed that the metric for the space-time has the form:

$$ds^2 = e^{\nu(r,t)}dt^2 - e^{\lambda(r,t)}dr^2 - r^2 d\theta^2 - r^2 \sin^2\theta d\varphi^2. \quad (2.48)$$

Calculating the corresponding Christoffel symbols, then the Riemann tensor, hence the Ricci tensor and the

Ricci scalar we obtain for the components of the Einstein tensor G^μ_ν:

$$G^t_t = e^{-\lambda}\left(\frac{\lambda'}{r} - \frac{1}{r^2}\right) + \frac{1}{r^2}, \tag{2.49}$$

$$G^r_t = -e^{-\lambda}\frac{\dot{\lambda}}{r} = -e^{\lambda-\nu}G^t_r, \tag{2.50}$$

$$G^r_r = -e^{-\lambda}\left(\frac{\nu'}{r} + \frac{1}{r^2}\right) + \frac{1}{r^2}, \tag{2.51}$$

$$G^\theta_\theta = G^\varphi_\varphi = \frac{e^{-\lambda}}{2}\left(\frac{\nu'\lambda'}{2} + \frac{\lambda'}{r} - \frac{\nu'}{r} - \frac{\nu'^2}{2} - \nu''\right) + \frac{e^{-\nu}}{2}\left(\ddot{\lambda} + \frac{\dot{\lambda}}{2} - \frac{\dot{\lambda}\dot{\nu}}{2}\right), \tag{2.52}$$

where the prime denotes differentiation with respect to r and the dot denotes differentiation with respect to t.

Then for the vacuum field equations

$$G^\mu_\nu = 0 \tag{2.53}$$

equations (2.49-2.52) give

$$e^{-\lambda}\left(\frac{\lambda'}{r}-\frac{1}{r^2}\right)+\frac{1}{r^2}=0, \tag{2.54}$$

$$e^{-\lambda}\left(\frac{\nu'}{r}+\frac{1}{r^2}\right)-\frac{1}{r^2}=0, \tag{2.55}$$

$$\dot{\lambda}=0. \tag{2.56}$$

The last equation shows that $\lambda = \lambda(r)$, and adding the first two equations (2.54) and (2.55) we get

$$\lambda+\nu=h(t), \tag{2.57}$$

where h is an arbitrary function of time. By a separation ansatz we can take $h = 0$, or later redefine the time coordinate so that the function h does not appear in the final solution. Thus we write

$$\lambda(r)=-\nu(r). \tag{2.58}$$

Now from equation (2.54) we have

$$e^{-\lambda}\left(1-r\lambda'\right)=1, \tag{2.59}$$

or equivalently

$$\frac{d}{dr}\left(re^{-\lambda}\right)=1. \tag{2.60}$$

Integration gives

$$re^{-\lambda} = const. + r. \tag{2.61}$$

We choose in view of Newtonian potential and dimensional consistenc: $const. = -2GM/c^2$, where M is the mass of the gravitational source. This gives

$$e^{-\lambda} = 1 - \frac{2GM}{c^2 r} = e^{\nu}. \tag{2.62}$$

Hence we have the *Schwarzschild solution*

$$\begin{aligned} ds^2 &= \left(1 - \frac{2GM}{c^2 r}\right) dt^2 - \left(1 - \frac{2GM}{c^2 r}\right)^{-1} dr^2 - \\ &\quad r^2 d\theta^2 - r^2 \sin^2\theta d\varphi^2. \end{aligned} \tag{2.63}$$

This solution represents the spacetime around a static, spherically symmetric gravitational source of mass M. It possesses a singularity at the origin $r = 0$, and at $r = 2GM/c^2$ called the *Schwarzschild singularity*.

The solution to the gravitational field equations, namely the metric components $g_{\alpha\beta}$, have been determined for various cases of interest. However the solution obtained by Schwarzschild for the case of a spherically symmetric spacetime and by Kerr for an axially symmetric spacetime, have proved to be of particular interest for various astrophysical phenomena such as for the gravitational field due to compact stellar objects. For their particular interest to us in the following chapters we outline these solutions here.

For a non-rotating, uncharged spherically symmetric star of mass M the spacetime is given by the Schwarzschild metric,

$$g_{\alpha\beta} = \begin{bmatrix} -(1-2M/r) & 0 & 0 & 0 \\ 0 & (1-2M/r)^{-1} & 0 & 0 \\ 0 & 0 & r^2 & 0 \\ 0 & 0 & 0 & r^2 sin^2\theta \end{bmatrix}. \tag{2.64}$$

For a rotating, uncharged, axially symmetric star of mass M, the metric has non-vanishing off diagonal components $g_{t\varphi} = g_{\varphi t}$ reflecting the conserved angular momentum of the star. For this case the spacetime is described by Kerr metric,

$$g_{\alpha\beta} = \begin{bmatrix} -(1-2Mr/\rho^2) & 0 & 0 & -2Marsin^2\theta/\rho^2 \\ 0 & \rho^2/\Delta & 0 & 0 \\ 0 & 0 & \rho^2 & 0 \\ -2Marsin^2\theta/\rho^2 & 0 & 0 & (r^2+a^2+\xi)sin^2\theta \end{bmatrix}, \tag{2.65}$$

where

$$\rho^2 \equiv r^2 + a^2cos^2\theta, \qquad \Delta \equiv r^2 + a^2 - 2Mr, \tag{2.66}$$

also $\xi = 2Mra^2sin^2\theta/\rho^2$, and the *Kerr parameter* $a \equiv J/M$ denotes the angular momentum of the rotating gravitational source per unit mass.

2.4 The Lense-Thirring Effect

The effects gravitational field due to a rotating compact star must be described in terms of the general theory of relativity. If the gravitational field is relatively weak, and rotational speed lies well below the relativistic limits, an approximation can be adequately used to linearize the field equations under the Lorentz gauge. The linearized equations, called the gravitoelectromagnetic field equations, posses striking formal similarities with the field equations of classical electromagnetic thcory. For instance analogous to the electromagnetic waves, the gravitoelectromagnetic equations predict the existence of gravitational waves. For a rotating gravitational source, the gravitoelectromagnetic theory leads to the existence of Lense-Thirring effect in terms of the gravitomagnetic force generated by the rotation of the mass (Lense & Thirring 1918). In fact any theory combining Newtonian gravity and Lorentz invariance consistently must include a gravitomagnetic force generated by mass current. The effect however is very small under normal terrestrial conditions. Attempts have been made to directly measure the Lense-Thirring frame drag due to Earth using laser ranging observations of the artificial satellites which gives an accuracy of 20% to 30% (Ciufolini, 1989, Lämmerzahl *et. al.* 2001). However the gravitomagnetically induced orbital precession of the artificial satellite is smaller than the effects caused by other perturbations such as tides and the quadrupole moment of the Earth. We now explore this analogy in more details in the following.

Generally equation (2.40) can be integrated to give

$$\zeta_{\alpha\beta} = -4\int\int\int \frac{T_{\alpha\beta}}{r} dV. \tag{2.67}$$

For a rotating mass, neglecting the stresses and the product of source velocities we have for the energy-momentum tensor

$$T_{\alpha\beta} = \rho u_\alpha u_\beta, \tag{2.68}$$

where $u_\alpha = (1, \mathbf{u})$. In component form we have

$$T_{\alpha\beta} = \rho \begin{bmatrix} 1 & u_1 & u_2 & u_3 \\ u_1 & 0 & 0 & 0 \\ u_2 & 0 & 0 & 0 \\ u_3 & 0 & 0 & 0 \end{bmatrix}. \tag{2.69}$$

It follows from the equation (2.67) and the above energy-momentum tensor that

$$\Phi = -\frac{1}{4}\zeta_{00} = -\int\int\int \frac{\rho}{r} dV, \tag{2.70}$$

and

$$\mathbf{A} = -\frac{1}{4}\zeta_{0i} = \int\int\int \frac{\rho\mathbf{u}}{r} dV, \tag{2.71}$$

where Φ is the Newtonian "gravitoelectric" potential and $\mathbf{A}$ is the "gravitomagnetic" potential.

Here it is evident that $T^{00} = \rho$ is the mass density (in gravitational units $c = 1 = G$), which for a homogeneous system of total mass M is given by

$$\iiint \rho dV = M, \tag{2.72}$$

thus $T^{i0} = J^i (\equiv \mathbf{J})$ represents the "mass-current" density, related to the total angular momentum of the system by

$$S^i = 2 \iiint \epsilon^i_{jk} x^j T^{k0} dV. \tag{2.73}$$

With the above definitions of the potentials we already notice a certain similarity with the classical electromagnetic theory. For instance the Lorentz gauge condition (2.35) can be written in terms of the potentials Φ and $\mathbf{A}$ as

$$\frac{\partial \Phi}{\partial t} + \mathbf{\nabla}.\mathbf{A} = 0, \tag{2.74}$$

the identification with the Lorentz gauge of the classical electromagnetic theory is striking. Moreover if we define, analogous to the classical electromagnetic theory the "gravitoelectric" and "gravitomagnetic" fields as

$$\mathbf{G} = -\boldsymbol{\nabla}\Phi, \quad \mathbf{H} = \boldsymbol{\nabla}\times 4\mathbf{A}. \tag{2.75}$$

respectively. It then follows from the gravitational field equations (2.67) and the Lorentz gauge condition that for a slowly rotating, weakly gravitating mass:

$$\boldsymbol{\nabla}\cdot\mathbf{G} = -4\pi\rho, \quad \boldsymbol{\nabla}\times\mathbf{G} = 0, \tag{2.76}$$

$$\boldsymbol{\nabla}\cdot\mathbf{H} = 0 \quad \boldsymbol{\nabla}\times\mathbf{H} = -4\pi\mathbf{J}. \tag{2.77}$$

We notice that in contrast to the classical electromagnetism, these equations involve a factor of 4. The origin of this factor lies in the fact that the above linear approximation involves a spin-2 field whereas in classical electrodynamics we deal with a spin-1 field.

Geodesic equation can be similarly reduced to a force law. Let us make the plausible assumption that the test particles in a compact gravitational source are not accreted to relativistic velocities, that is, the velocity of light. Then the line element ds is identical to dt. Then for the geodesic equation we have

$$\frac{d^2x^i}{dt^2} = -\Gamma^i{}_{\alpha\beta}\frac{dx^\alpha}{dt}\frac{dx^\beta}{dt}. \tag{2.78}$$

For the Christoffel symbol we substitute from (2.25) and (2.28) into (2.78), then the geodesic equation becomes

$$\frac{d^2x^i}{dt^2} = -(\zeta^i_{\alpha,\beta} - \frac{1}{2}\zeta^i_{\alpha\beta,} - \frac{1}{4}\eta^i_\alpha\zeta_{,\beta} - \frac{1}{4}\eta^i_\beta\zeta_{,\alpha} + \frac{1}{4}\eta_{\alpha\beta}\zeta_{,}{}^i)\frac{dx^\alpha}{dt}\frac{dx^\beta}{dt}, \tag{2.79}$$

where $\zeta = \eta^{\alpha\beta}\zeta_{\alpha\beta} = -h$. Further more $\zeta = \zeta_{00}$. Denote

$$\frac{dx^\alpha}{dt} \equiv u^\alpha, \tag{2.80}$$

we have $u^\alpha = (1, u^i)$. Then the geodesic equation (2.79) reduces to

$$\frac{d^2x^i}{dt^2} = -\zeta^i_{0,j}u^j + \zeta_{j0,}{}^i u^j + \frac{1}{4}\zeta_{00,}{}^i. \tag{2.81}$$

As before we denote

$$\Phi = -\frac{1}{4}\zeta_{00}, \tag{2.82}$$

and

$$\mathbf{A} = -\frac{1}{4}\zeta^{i}_{\ 0}. \tag{2.83}$$

Then the geodesic equation becomes

$$\frac{d^2\mathbf{r}}{dt^2} = \mathbf{grad}\Phi - \mathbf{u} \times \mathbf{curl}\ \mathbf{A}. \tag{2.84}$$

We further use for $\mathbf{G}$ and $\mathbf{H}$ in (2.75) to obtain the gravitational force law

$$m\frac{d^2\mathbf{r}}{dt^2} = m(\mathbf{G} + \mathbf{v} \times \mathbf{H}). \tag{2.85}$$

where we have assumed the identity of gravitational and inertial mass of the test mass. Again the formal similarity with the classical electromagnetic theory is striking. The only differences are: the minus sign in (2.76, 2.77), due to the attractive nature of the gravitational force; and the novel factor 1/4 in (2.82). Due to its formal analogy to the classical electromagnetic theory, the above approximation is referred to as the gravitoelectromagnetic approximation to the general theory of relativity.

This leads to the Lense-Thirring Effect. We consider the gravitational field inside a rotating spherical shell of mass M, radius R, and angular velocity $\mathbf{\Omega}$. As in classical electromagnetic theory, the gravitoelectric force (analogous to the electric field) inside such a shell of mass M vanishes, whereas the gravitomagnetic force (analogous to the magnetic force) is given by

$$\mathbf{H} = \frac{2}{3}\frac{M\mathbf{\Omega}}{R}, \tag{2.86}$$

correspondingly a test particle of unit mass moving with a velocity $\mathbf{v}$ experiences a force

$$\mathbf{F}_g = \frac{2M}{3R}\mathbf{v} \times \mathbf{\Omega}, \tag{2.87}$$

which is the Coriolis force of classical mechanics.

For practical purposes the graviomagnetic field acting on a test mass in vicinity of the Earth can be approximated by a dipolar field. This corresponds to the case of a charged particle orbiting about a magnetic dipole. The gravitomagnetic approximation then gives for the angular velocity precession (Ruggiero &

Tartaglia 2002):

$$\text{precession rate} = (\frac{G}{c^2})\frac{3\mathbf{r}(\mathbf{r}.\mathbf{S}) - \mathbf{S}r^2}{r^5}. \quad (2.88)$$

For the Earth this precession has value about 0.05 $arcsec/yr$. As pointed out earlier the effects, such as due to the quadrupole moment of the Earth, are of greater magnitude than gravitomagnetic ones. Recently, however, data obtained from laser ranging satellites LAGEOS and LAGEOS II had disclosed the gravitomagnetically induced precession within 20 to 30 per cent accuracy. In astrophysics, observational evidence for the Lense-Thirring frame dragging has been detected for the periastron precession rate of the binary pulsar PSR 1913+16 (Will 2001).

For the case of a slowly rotating gravitational mass such as a compact star the gravitoelectromagnetic analogy to the magnetic dipole is not generally applicable. Here for the star, not rotating very fast, the effects of gravitoelectric force are dominant and hence to a high degree of approximation the star can be regarded as a rotating spherical mass M with uniform mass density. Then analogous to the case of slowly rotating charged sphere in classical electromagnetism, we have for the gravitoelectric force (Rindler 2006)

$$\mathbf{G} = -M\hat{\mathbf{r}}/r^2, \tag{2.89}$$

and for the gravitomagnetic

$$\mathbf{H} = -\frac{12}{5}MR^2(\mathbf{\Omega}.\mathbf{r}\frac{\mathbf{r}}{r^5} - \frac{1}{3}\frac{\mathbf{\Omega}}{r^3}). \tag{2.90}$$

This result was first given by Thirring and Lense (Lense & Thirring 1918).

Chapter 3

Maxwell Equations

3.1 Maxwell Equations in SR

In special relativity (SR), Minkowski geometry naturally leads to the invariance of Maxwell field equations under a linear transformation of coordinates, that is, Maxwell equations possessed a tensor character under the Lorentz transformation. By defining the *electromagnetic field tensor*,

$$F_{\mu\nu} \equiv -A_{[\mu,\nu]} = A_{\nu,\mu} - A_{\mu,\nu}\,, \tag{3.1}$$

where $(A_\mu) = (\phi, \mathbf{A}) = (\phi, A_x, A_y, A_z)$ is the four component electromagnetic potential, Maxwell equations can be written in a covariant form:

$$F_{\mu\nu},_\nu = 4\pi J_\mu, \tag{3.2}$$

where J_μ is the current four-vector: $J_\mu \equiv \rho v_\mu = (\rho c, \mathbf{J})$ with ρ as charge density and $\mathbf{J}$ the current vector in $\mathbf{R}^3$. Here $v_\mu = (c, \mathbf{v})$ is velocity 4-vector. The condition

$$J_\mu,_\mu = 0, \tag{3.3}$$

expresses the charge continuity equation

$$\frac{\partial \rho}{\partial t} + \boldsymbol{\nabla} \cdot \mathbf{J} = 0. \tag{3.4}$$

Further the Lorentz force law can be written in a covariant form

$$K_\mu = q v^\lambda F_{\lambda\mu}, \tag{3.5}$$

where K_μ is the force 4-vector.

The metric tensor $\eta_{\alpha\beta}$ is the flat Minkowski spacetime metric. It directly relates the properties of space (spacetime) to the kinematical properties of moving particles, since special theory applies to uniform motion only. The spacetime metric when there is a non-uniform motion (acceleration), such as in the case of a gravitational field,can be written locally as

$$ds^2 = g_{\alpha\beta} dx^\alpha dx^\beta. \tag{3.6}$$

When $g_{\alpha\beta}$ are interpreted as the potentials of a gravitational field then these represent solutions to the Einstein field equations:

$$R_{\alpha\beta} - \frac{1}{2} g_{\alpha\beta} R = \kappa T_{\alpha\beta}, \tag{3.7}$$

where $R_{\alpha\beta}$ is the contracted Riemann tensor, R is the invariant obtained by the further contraction of $R_{\alpha\beta}$, $T_{\alpha\beta}$ is the energy-momentum tensor and κ is a constant of proportionality. This fact enables us to correlate the gravitational field equations to the Maxwell equations for the electromagnetic field. It must be emphasized here that the field potentials play a very important role in this method of unification. The general procedure is to solve the gravitational field equations for a given space-time thus to obtain explicit expressions for $g_{\alpha\beta}$. Then by using the expression (3.2) for the electromagnetic field tensor in general covariant form, we obtain the Maxwell's equations in a curved spacetime. The general form of the electromagnetic field tensor for the generalized (covariant) Maxwell equations is obtained in the following section.

3.2 Maxwell Equations in GR

In general relativity (GR) Maxwell equations can be obtained as follows. Let $F_{\alpha\beta}$ be a 2-form *i.e.*, a skew-symmetric tensor field of order two. In our case $F_{\alpha\beta}$ represents the electromagnetic field tensor. Let $(^*F)_{\alpha\beta}$ be its dual in a spacetime V_4 where we have chosen an orientation. If $\eta_{\alpha\beta\gamma\delta}$ is the volume 4-form of V_4 corresponding to this orientation then we can write

$$(^*F)_{\alpha\beta} = \frac{1}{2}\eta_{\alpha\beta\gamma\delta}F^{\alpha\beta}. \tag{3.8}$$

Note that if we consider a p-form on a Riemannian manifold V_n then we have

$$*^2 = \frac{g}{\mid g \mid}(-1)^{p(n-p)}. \tag{3.9}$$

Here with our signature for V_4 we have for the 2- forms $*^2 = -1$. Thus

$$({}^{**}F)_{\alpha\beta} = -F_{\alpha\beta}. \tag{3.10}$$

For u^α a unit four-vector at event $\mathbf{x}\epsilon V_4$ defining a time like direction at this point, we define

$$E_\beta = u^\alpha F_{\alpha\beta}, \tag{3.11}$$

and

$$B_\beta = u^\alpha ({}^*F)_{\alpha\beta}. \tag{3.12}$$

It is clear that there 4-vector (E_β, B_β) are orthogonal to u^α i.e.

$$u^\alpha E_\alpha = u^\alpha B_\alpha = 0. \tag{3.13}$$

Now let us consider at $\mathbf{x}\epsilon V_4$ an orthogonalized frame which is a rest frame with respect to u^α. In this frame the components E_0 and B_0 are zero and we have:

$$E_i = F_{0i}, \quad B_i = ({}^*F)_{0i}, \tag{3.14}$$

where the Latin suffix i takes the values $1, 2,$ and 3. It follows that

$$E^i = F^{0i}, \quad B^i = ({}^*F)^{0i}. \tag{3.15}$$

Since we have

$$({}^*F)_{0i} = \frac{1}{2}\eta_{0ijk}F^{jk}, \tag{3.16}$$

therefore

$$F_{0i} = -\frac{1}{2}\eta_{0ijk}(^*F)^{jk}. \tag{3.17}$$

In the rest frame where $u^o = 1$ and $u^i = 0$ (time-like), we can write $E_i = F_{0i}$ as

$$F_{0i} = u_0 E_i - u_i E_0 - \eta_{0ijk} u^i B^k. \tag{3.18}$$

Also from equations (3.16) and (3.17) we get

$$F_{jk} = -\eta_{jk0i}(^*F)^{0i}. \tag{3.19}$$

Using equation (3.15) we have

$$F_{jk} = -\eta_{jk0i} B^i. \tag{3.20}$$

This can be put in the form

$$F_{jk} = u_j E_k - u_k E_j - \eta_{ij\lambda\mu} u^\lambda B^\mu, \tag{3.21}$$

where i, j, k range from 1 to 3; and $\lambda, \mu = 0, 1, 2, 3$. With the help of equation (3.18) and (3.21) we arrive at the unique tensorial expression,

$$F_{\alpha\beta} = u_\alpha E_\beta - u_\beta E_\alpha - \eta_{\alpha\beta\lambda\mu} u^\lambda B^\mu, \tag{3.22}$$

or

$$F_{\alpha\beta} = u_{[\alpha}E_{\beta]} - \eta_{\alpha\beta\lambda\mu}u^{\lambda}B^{\mu}. \tag{3.23}$$

Expression (3.23) gives the electromagnetic field tensor in a covariant form. The volume element 4-form of V_4 namely $\eta_{\alpha\beta\lambda\mu}$ is defined as

$$\eta_{\alpha\beta\lambda\mu} = \sqrt{-g}\epsilon_{\alpha\beta\lambda\mu}, \tag{3.24}$$

where g represents the determinant of the metric tensor $g_{\alpha\beta}$ and $\epsilon_{\alpha\beta\lambda\mu}$ is the Levi -Civita symbol, defined as

$$\begin{aligned} \epsilon_{\alpha\beta\gamma\delta} &= +1, \quad \text{for a cyclic permutation of } \alpha\beta\gamma\delta; \\ &= -1, \quad \text{for an anticyclic permutation of } \alpha\beta\gamma\delta; \\ &= 0, \quad \text{if } (\alpha\beta\gamma\delta) \text{ is not a permutation.} \end{aligned} \tag{3.25}$$

The electromagnetic field tensor can also be expressed in a contravariant form as follows. From equation (3.23) we have

$$F_{\alpha\beta} = 2u_{[\alpha}E_{\beta]} + \eta_{\alpha\beta\gamma\delta}u^{\gamma}B^{\delta}, \tag{3.26}$$

which gives

$$F^{ab} = g^{a\alpha}g^{b\beta}F_{\alpha\beta}, \tag{3.27}$$

that is

$$F^{ab} = g^{a\alpha}g^{b\beta}(u_\alpha E_\beta - u_\beta E_\alpha) + g^{a\alpha}g^{b\beta}\eta_{\alpha\beta\gamma\delta}u^\gamma B^\delta, \tag{3.28}$$

or

$$F^{ab} = (g^{a\alpha}u_\alpha g^{b\beta}E_\beta - g^{b\beta}u_\beta g^{a\alpha}E_\alpha) + g^{a\alpha}g^{b\beta}\eta_{\alpha\beta\gamma\delta}u^\gamma B^\delta, \tag{3.29}$$

or

$$F^{ab} = (u^a E^b - u^b E^a) + \eta^{ab}_{\gamma\delta}u^\gamma B^\delta. \tag{3.30}$$

Thus we get the contravariant form of the metric tensor

$$F^{\alpha\beta} = \frac{1}{2}u^{[\alpha}E^{\beta]} + \eta^{\alpha\beta\gamma\delta}u_\gamma B_\delta\ . \tag{3.31}$$

It should be noted that the choice of spacetime V_4 is quite general here, namely a four dimensional vector space, or more generally, a differentiable manifold of dimension four; provided that the velocity four vector is chosen consistently. Also, the signature of the any metric defined on this manifold must be either $(+ - --)$ or $(- + ++)$.

A physically useful class of observers consists of those who remain stationary with respect to infinity. For such observers no spatial coordinate r, θ, or φ should change with time along the world line. To do so some artificial means of power generation (such as a rocket thrust) had to be used to sustain their position in space. Although in a Schwarzchild spacetime such an observer can get arbitrarily close to the spacetime horizon (that is where the metric component g_{tt} vanishes), this is not true in the Kerr spacetime. To see this we first note that for a stationary observer the four velocity is the time component, $u^t = dt/d\tau$. Therefore using $u^\alpha u_\alpha = -1$, we have

$$g_{tt}(u^t)^2 = -1, \tag{3.32}$$

which determines u^t. For Schwarzschild spacetime g_{tt} remains well defined at a distance arbitrarily close to the horizon. However for the Kerr spacetime there is a surface at

$$r \equiv r_e(\theta) = M + \sqrt{M^2 - a^2 cos^2\theta}, \tag{3.33}$$

enveloping the event horizon, where the metric component g_{tt} vanishes. At this surface, called the static limit, does not hold. We notice that in the region between the static limit and the horizon, called the ergosphere, the metric component g_{tt} is positive; therefore at the static limit and inside the ergosphere no stationary observer with a timelike four velocity is possible. There is however another physically meaningful choice of observers. Although it is not possible for an observer to remain fixed at r, θ, and φ; it is possible to define an observer remaining fixed at r, and θ but not at φ. Such an observer co-rotates with the star with respect to infinity, and hence is called a ZAMO (**Z**ero **A**ngular **M**omentum **O**bserver). For such an observer $u^{\varphi} \neq 0$ implies that

$$u^{\varphi} = \frac{d\varphi}{dt}\frac{dt}{d\tau} = \omega u^{t}, \tag{3.34}$$

where

$$\omega \equiv \frac{d\varphi}{dt} = \frac{u^{\varphi}}{u^{t}}. \tag{3.35}$$

From the last expression we obtain for ω

$$\omega = \frac{g^{t\varphi}}{g^{tt}} = \frac{2Mra}{(r^2+a^2)^2 - \Delta a^2 sin^2\theta} = \omega(r,\theta). \tag{3.36}$$

Furthermore using $u^\alpha u_\alpha = -1$, we have for a ZAMO

$$u^t = 1/\sqrt{-(g_{tt} + \omega g_{t\varphi} + \omega^2 g_{\varphi\varphi})}. \tag{3.37}$$

In the slow rotation approximation we can neglect the terms containing ω^2 and higher (correspondingly a^2 and higher). In this approximation the θ dependence for both ω and u^t is lost, and we obtain for a ZAMO

$$\omega = \frac{2Ma}{r}, \tag{3.38}$$

and

$$u^t = 1/\sqrt{-(g_{tt})} = 1/\sqrt{1 - \frac{2M}{r}}. \tag{3.39}$$

Clearly a ZAMO is a useful class of observers for astrophysical observations. Physically these observers circle the gravitational source with a given angular velocity at a fixed radial distance and polar angle.

Example: The set of Einstein-Maxwell equations

$$F_{\alpha\beta,\gamma} + F_{\beta\gamma,\alpha} + F_{\gamma\alpha,\beta} = 0, \tag{3.40}$$

reduces

$$(\sqrt{-g}u^t B^r)_{,r} + (\sqrt{-g}u^t B^\theta)_{,\theta} + (\sqrt{-g}u^t B^\varphi)_{,\varphi} = 0, \quad (3.41)$$

$$(\sqrt{-g}u^\varphi B^\theta)_{,\theta} + (\sqrt{-g}u^t B^\varphi)_{,t} + (\sqrt{-g}u^\varphi B^r)_{,r} = 0, \quad (3.42)$$

$$(\sqrt{-g}u^\varphi B^\theta)_{,\varphi} + (\sqrt{-g}u^t B^\theta)_{,t} = 0, \quad (3.43)$$

$$(\sqrt{-g}u^\varphi B^r)_{,\varphi} + (\sqrt{-g}u^t B^r)_{,t} = 0, \quad (3.44)$$

which can be further simplified into

$$\frac{1}{u^t}(\sqrt{-g}u^t B^r)_{,r} + (\sqrt{-g}B^\theta)_{,\theta} + \sqrt{-g}B^\varphi{}_{,\varphi} = 0, \quad (3.45)$$

$$\sqrt{-g}B^\varphi{}_{,t} + \frac{1}{u^t}(\sqrt{-g}\omega u^t B^r)_{,r} + \omega(\sqrt{-g}B^\theta)_{,\theta} = 0, \quad (3.46)$$

$$B^\theta_{,t} + \omega B^\theta{}_{,\varphi} = 0, \quad (3.47)$$

$$B^r{}_{,t} + \omega B^r_{,\varphi} = 0, \quad (3.48)$$

where ω is given by (3.38) and u^t is given by (3.39).

Differentiating with respect to r the second term in equation (3.45) as a product of $\omega(r)$ and $\sqrt{-g}u^t B^r$, and using equation (3.46) in the resulting expression we obtain a single partial differential equation

$$B^{\varphi}{}_{,t} + \omega B^{\varphi}{}_{,\varphi} = -B^{r}\omega_{,r}. \tag{3.49}$$

Now to solve equation (3.47), (3.48), and (3.49) we apply the method of characteristics. Suppose $f(t,\varphi)$ represents any of the component $(B^r, B^\theta, B^\varphi)$, then we have

$$\frac{df}{dt} = f_{,t} + f_{,\varphi}\frac{d\varphi}{dt}. \tag{3.45}$$

Comparing equation (3.45) with (3.49), and setting $f = B^\varphi$ we obtain two simultaneous ordinary differential equations

$$\frac{dB^{\varphi}}{dt} = -B^{r}\omega_{,r}, \tag{3.46}$$

and notably

$$\frac{d\varphi}{dt} = \omega. \tag{3.47}$$

Equation (3.47) offers an interesting interpretation. This relation implies that the characteristics travel at a speed ω. Solving (3.47) we get the characteristics $\varphi = \omega t + \varphi_0$. On the other hand equation (3.46) has solution $B^\varphi = -B^r(\varphi_0)t\omega_{,r} + h(\varphi_0)$. Combining the two results we have the general solution to equation (3.45),

$$B^{\varphi} = -B^{r}(\varphi - \omega t)t\omega_{,r} + h(\varphi - \omega t), \tag{3.48}$$

where the function $h(t, \varphi)$ satisfies the condition

$$h_{,t} + \omega h_{,\varphi} = 0. \tag{3.49}$$

A similar procedure for equations (3.47) and (3.48) shows that B^r and B^θ indeed have general solutions of the form $f(\varphi - \omega t)$, where f is an arbitrary function of its argument $\varphi - \omega t$, and whose form is determined by the physical constraints.

The magnetic field outside a slow rotating gravitational source has a general travelling wave solution of the type

$$\begin{aligned} B^{r} &= f(\varphi - \omega t), \quad B^{\theta} = g(\varphi - \omega t), \\ B^{\varphi} &= -ft\omega_{,r} + h(\varphi - \omega t), \end{aligned} \tag{3.50}$$

where f, g, and h are arbitrary functions of the argument $\varphi - \omega t$ and where ω given by (3.47) can be interpreted as the velocity with which a characteristic moves or the velocity of a wave profile. To further clarify the physical significance of these solutions, we discuss some specific cases of f, g, and h and their dependence on ω.

If the magnetic field remains stationary outside the source, then it follows from (3.50) that the magnetic field components are given by

$$B^r = \varphi - \omega t, \quad B^\theta = \varphi - \omega t, \quad B^\varphi = -(\varphi - \omega t)t\omega_{,r}, \tag{3.51}$$

where for simplicity we have taken $h = 0$. Furthermore φ being the azimuthal angle can be given a fixed value, say zero. We notice that for a fixed value of φ, the magnetic field components depend directly on the frame dragging frequency ω. This means that the magnetic field, in this case, is induced by the rotation of the Kerr black hole. However this induced field may not last for long as the assumption of being stationary is not valid for actual magnetospheres.

For an oscillating magnetic field the functions f, g, and h have a sinusoidal form represented by sin or cos functions. As before we take $h = 0$, and $\varphi = 0$. Then for the magnetic field components we have from (3.50)

$$B^r = \cos\omega t, \quad B^\theta = \sin\omega t, \quad B^\varphi = -t\omega_{,r}\cos\omega t. \tag{3.52}$$

Here we notice the frequency of oscillations is the dragging frequency. The magnetic field in this case is again induced by rotating gravitational source. We notice that the field components B^r and B^θ remain bounded for all time, however the component B^φ is unbounded as t increases. For actual physical situations the magnetic field must decay with the passage of time. Again the oscillatory solutions are not valid for long periods time, for a real physical situation.

The solutions (3.52) to the Einstein-Maxwell equation, for $h = 0$ and $\varphi = 0$, can be written in the form of exponentially decaying functions as

$$\begin{aligned} B^r &= \exp(-\omega t), \quad B^\theta = \exp(-\omega t), \\ B^\varphi &= -t\omega_{,r}\exp(-\omega t), \end{aligned} \tag{3.53}$$

or in terms of complex representation

$$
\begin{aligned}
B^r &= \exp(-i\omega t), \quad B^\theta = \exp(-i\omega t), \\
B^\varphi &= -t\omega_{,r}\exp(-i\omega t).
\end{aligned} \tag{3.54}
$$

Both of these solutions are unbounded for very large time scales and each cannot be used to represent physically the magnetic field.

The above examples show that the magnetic field, for a given value of φ, is induced by the rotating gravitational source and it can decrease with distance from the source. Furthermore the magnetic field is unbounded in these cases. We can interpret this as follows: *For a gravitational source with a magnetic field there are magnetic waves travelling along the azimuthal with velocity equal to the angular velocity of a free falling inertial frame induced by the rotating gravitational source and which remain unbounded function of time in general.* In the following, however, it is shown that a bounded physical solution exits for the magnetic field also.

3.3 Toroidal Field Solution

If we interpret $g_{\alpha\beta}$ as the potentials of a gravitational field determined by the Einstein field equations then we can write the electromagnetic field equations in a spacetime whose geometry is given by the general relativistic field equations. The resulting set of equations, called the generalized Maxwell field equations, can be expressed as

$$F_{\alpha\beta,\gamma} + F_{\beta\gamma,\alpha} + F_{\gamma\alpha,\beta} = 0, \tag{3.55}$$

$$\left(\sqrt{-g}F^{\alpha\beta}\right)_{,\beta} = 4\pi\sqrt{-g}J^{\alpha}, \tag{3.56}$$

where the electromagnetic field tensor, in covariant and contravariant form, is given by expressions (3.23) and (3.31). Also here, as usual, $,\alpha$ as a subscript denotes the ordinary partial derivative with respect to the coordinate x^{α}.To discuss these equations in full we must choose the four velocity vector, that is the co-moving observer has to be specified. As described above, a convenient and physically meaningful choice of the co-moving observer is the one belonging to the class of observers having zero angular momentum (ZAMO).

Such an observer, circling the Kerr black hole with angular velocity ω at a fixed r and θ, is brought into motion by the dragging of the background spacetime. Thus for a ZAMO u_r and u_θ vanish and the velocity four vector is given by $(u^t, 0, 0, u^\varphi)$. Moreover, since the ZAMO defines a co-moving frame therefore, the electric field vanishes also. Under these conditions we obtain the following set of generalized Maxwell field equations for the first pair (3.55),

$$(\sqrt{-g}u^t B^r)_{,r} + (\sqrt{-g}u^t B^\theta)_{,\theta} + (\sqrt{-g}u^t B^\varphi)_{,\varphi} = 0, \tag{3.57}$$

$$(\sqrt{-g}u^\varphi B^\theta)_{,\theta} + (\sqrt{-g}u^t B^\varphi)_{,t} + (\sqrt{-g}u^\varphi B^r)_{,r} = 0, \tag{3.58}$$

$$(\sqrt{-g}u^\varphi B^\theta)_{,\varphi} + (\sqrt{-g}u^t B^\theta)_{,t} = 0, \tag{3.59}$$

$$(\sqrt{-g}u^\varphi B^r)_{,\varphi} + (\sqrt{-g}u^t B^r)_{,t} = 0, \tag{3.60}$$

and similarly from expression (3.56) the second pair gives,

$$(u_\varphi B_\theta)_{,r} + (u_\varphi B_r)_{,\theta} = 4\pi\sqrt{-g}J^t, \tag{3.61}$$

$$-(u_\varphi B_\theta)_{,t} + (u_t B_\varphi)_{,\theta} - (u_t B_\theta)_{,\varphi} = 4\pi\sqrt{-g}J^r, \tag{3.62}$$

$$-(u_\varphi B_r)_{,t} - (u_t B_\varphi)_{,r} + (u_t B_r)_{,\varphi} = 4\pi\sqrt{-g}J^\theta, \quad (3.63)$$

$$(u_t B_\theta)_{,r} - (u_t B_r)_{,\theta} = 4\pi\sqrt{-g}J^\varphi. \quad (3.64)$$

Given the current four vector J^α, the metric tensor $g_{\alpha\beta}$, and the velocity four vector u^α the set of equations (3.57)-(3.60) can be, to give the magnetic field as a function of (t, r, θ, φ). For a rotation source mass M the dragging frequency is,

$$\omega(r) \equiv \frac{d\varphi}{dt} = \frac{2GJ}{c^2 r^3}, \quad (3.65)$$

which equals the angular velocity of a free falling inertial frame, whereas J is the total angular momentum of the Kerr black hole as measured from infinity. Finally, under the assumption of slow rotation (i.e., $\omega(r)^2$ and higher powers are neglected), we take explicitly the velocity four vector in ZAMO as follows:

$$u^\alpha = (u^t, 0, 0, \omega u^t), \quad u_\alpha = (-\frac{1}{u^t}, 0, 0, 0). \quad (3.66)$$

The toroidal field induced spacetime dragging can be derived from the coupling of the magnetic field to the gravitational field of a massive source. In this the energy density due to the toroidal field contributes to the total energy of a test object. For this we consider the Maxwell equations in curved spacetime around a rotating mass (Mirza 2017; Pétri 2013; Mirza 2007; Oron 2002). The background space in this case is axially symmetric, and for slowly rotating objects is given by the linearized Kerr metric. It can be easily shown that the assumption of an everywhere finite J^α leads to the condition that in a co-moving frame $E^\alpha = 0 = E_\alpha$ (Lichnerowicz 1967). Therefore for the exterior region to the star, the electromagnetic field tensor takes the form $F_{\alpha\beta} = \sqrt{-g}\epsilon_{\alpha\beta\gamma\delta}u^\gamma B^\delta$,and in the contravariant form as $F^{\alpha\beta} = -(-g)^{-1/2}\epsilon^{\alpha\beta\gamma\delta}u_\gamma B_\delta$,where $\epsilon_{\alpha\beta\gamma\delta}$ is the four index Levi -Civita symbol.

Maxwell equations (3.57)-(3.64) can be solved to give the magnetic field as a function of (t, r, θ, φ). It forms a system of coupled partial differential equations. Under the slow rotation condition it is found (Mirza 2007) that for $\omega \neq 0$ and $B_t \neq 0$, the magnetic field around the rotating mass is,

$$(B^\alpha) = B_0 \left(0, 0, \frac{A(\theta)\sin\theta\sin\xi}{u_t r^2 \sin^2\theta}, \frac{\cos\xi}{u_t r^2 \sin^2\theta} \right), \quad (3.67)$$

where B_0 is a constant, $\xi = \varphi - \omega t$. Here

$$A(\theta) = \int \frac{d\theta}{\sin\theta}. \tag{3.68}$$

According to equations (3.67) and (3.68), the contribution of the poloidal field to the total field energy density for any close loop for θ vanishes. The toroidal field, however, contributes to the energy density of the surrounding field for a rotating source, by

$$\epsilon_\varphi = \frac{B_0^2 \cos^2\xi}{u_t^2 r^2 \sin^2\theta}. \tag{3.69}$$

The field energy density is minimum in the equatorial plane $\theta = \pi/2$ and maximum at the poles. This is significant for the orbital clustering, since the toroidal field energy density provides the outward velocity drag (Figure 3.1).

To calculate the velocity drag induced by the toroidal field, we note that the dragging effects are maximum in the equatorial plane $\theta = \pi/2$ of the star, we therefore consider the toroidal field component explicitly as,

$$B_\varphi = -\frac{B_0}{r\sqrt{1 - \frac{R_s}{r}}} \cos(\varphi - \omega t), \tag{3.70}$$

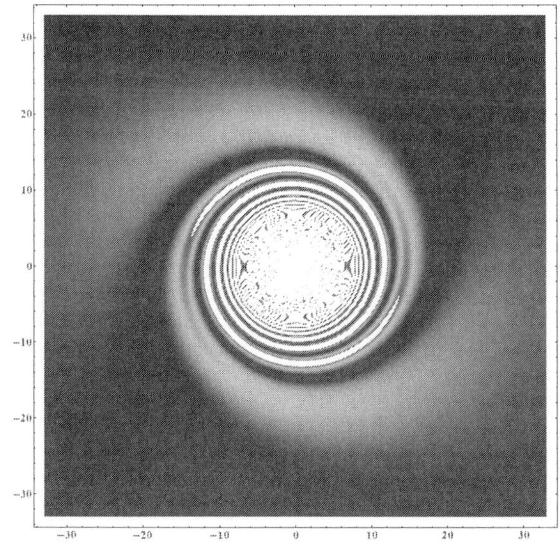

Figure 3.1: Magnetic field energy distribution (3.69). The spiral arms correspond to the regions where magnetic field energy density is higher due to the magnetogravitational dragging. The illustration shows matter (energy) compressed under the magnetogravitational field.

where $r > R_s$, and $R_s = 2GM/c^2$ is the Schwarzschild radius.

For a slowly rotating star like the Sun $\omega << 1$, and the cosine factor depends on the choice of the coordinate φ, hence can be absorbed in the constant B_0. The radial drag therefore comes from the factor $B_0/\left(r\sqrt{1 - R_s/r}\right)$. If R denotes the radius of the disc in the ecliptic as the equatorial plane of the star, containing total mass M, then the *total* energy density of the field at radius R is,

$$\epsilon = \int_0^R 4\pi r^2 \epsilon_\varphi dr. \tag{3.71}$$

Substituting from equation (3.70), and integrating by parts we obtain,

$$\epsilon = 4\pi B_0^2 \left[R + R_s \ln\left(R - R_s\right)\right] \cos^2(\varphi - \omega(R)t), \; R > R_s, \tag{3.72}$$

where ω^2 and higher order powers have been neglected. We see that this energy contributes to the total energy of the system, hence like the gravitational potential energy is independent of the mass of the test body. The second term in equation (3.72) is very small as compared to the first, hence can be neglected. The energy density due to the toroidal-gravitational field coupling therefore is,

$$\epsilon = 4\pi B_0^2 R \cos^2(\varphi - \omega(R)t), \tag{3.73}$$

depending on the frame-dragging frequency ω. In gravitationally bound systems, the virial theorem applies. For potential energy $V(R)$ and kinetic energy $K = v^2/2$ (per unit mass), we have by the virial theorem, $2K = V = \epsilon$. This gives for the dragging velocity v, the formula,

$$v = B_0\sqrt{4\pi R}\cos(\varphi - \omega(R)t). \tag{3.74}$$

This gives the outward (positive) drag velocity for a test body in the gravitational field of the sun, in addition to its orbital velocity induced by the Newtonian gravitational potential. Here (in equation (3.74)), in accordance with the equivalence principle, the induced velocity is independent of the test mass.

The above demonstrates a coupling between the magnetic field and the gravitational field of a massive object. The coupled field is called the *magneto-gravitational field*. Such a coupling exists in the form of outward moving density waves, expanding in a double arm spiral.

Chapter 4

Galactic Dynamics

4.1 Velocity Curves

Rotation curves give the velocity of galactic mass as a function of distance from the galactic center. The velocity curve exhibit non-Keplerian orbital motion. In a Keplerian orbit, the velocity of the test mass decreases sharply over the distance from the source mass, so that at the edge of the galactic disk the velocity is less than a few kilometers per second. The observed behavior of galaxies, especially in the spiral galaxies is very different. Instead of decreasing with distance from the galactic center, the velocity either remains almost uniform in the disk, or actually increases. This effect was first noticed by F. Zwicky (1937) and also indicated by J. Oort (1932) during the 1930s. Later, V. Rubin (1983) and others (see, Sofue & Rubin 2001, and references therein) found better evidence to support Zwicky's hypothesis of missing galactic mass. Zwicky's reasoning was however based on non-luminous gravitational sources, which turned out be incorrect due to the lack of such objects in the galactic disk. The idea has been replaced with weakly interacting matter (dark matter) to account for the galactic rotational velocity curves

Here the gravitomagnetic coupling is used to deduce the non-Keplerian behavior of the galactic rotation. As described in the previous chapters, the coupled field induces rotational dragging at large distances. Farther away from the central gravitational source this effect dominates the orbital motion. We deduce here not only the velocity curves for a number of different galaxies, but also relate them to other observed features of galaxies. These include the Tully-Fisher formula, which relates the total galactic mass to the maximum velocity in the galaxy, and the age of a galaxy based on its size and magnetic field. A simple formula is also give for the size of a galaxy when its rotational speed is known. These results are compared with the observed galactic parameters.

In Chapter 3, we have found that the magnetic field around a rotating compact astrophysical source can be derive from Maxwell equations in slowly rotating Kerr spacetime (Mirza 2007, 2017). Using the standard spacetime coordinates, (r, θ, φ, t), where the azimuthal angle φ is measured in the galactic plane, the toroidal magnetic field in the galactic disk $(\theta = \pi/2)$ is given by,

$$B(r) = \frac{B_0 \cos\xi}{r u_t}, 0 < \xi \leq \pi, \tag{4.1}$$

where $u_t = -\sqrt{1 - r_s/r}$, $r_s = 2GM_\bullet/c^2$ being the the Schwarzschild radius, and M_c is the mass of the compact source in the galactic center. This field is coupled to the spacetime vorticity via $\xi = \varphi - \omega t$, where $\omega \neq 0$ is the Lense-Thirring dragging frequency. For a slowly rotating mass, $\omega(r) = 2J/r^3$, where J is the angular momentum of the central compact source mass M_c.

In the equatorial plane identified as the galactic plane, the total energy density due to the magnetic field interior to a circular orbit of radius r around the source at origin can be calculated by first integrating over the variable ξ, which gives a constant contribution absorbed in B_0. Therefore, the magnetic energy density at any point r in the galactic disk is given by,

$$\epsilon_B = \int_0^r 4\pi r'^2 B^2 dr'. \tag{4.2}$$

Putting from equation (4.1) into equation (4.2), we obtain,

$$\epsilon_B = 4\pi B_0^2 \left[r + r_s \ln(r - r_s) \right], \quad r > r_s. \tag{4.3}$$

Equation (4.3) gives the energy density due to the magnetic field coupled to the spacetime dragging, also the total energy has contribution due to the Newtonian gravitational potential energy for the net mass M contained within the orbit. Correspondingly, in the outer region $r > r_s$, disk velocity has contribution v_0 due to the source at $r = r_s$, the velocity field v in the exterior due to the magnetic energy, and the Keplerian velocity due to the gravitational potential energy. The effective potential energy of gravity per unit mass is $E_G = GM/r$ where M is the total mass contained within the orbit. Defining the dimensionless magnetic energy density $E_B = \epsilon_B/B_0^2$, we equate the net potential energy $(E_B + E_G)$ to the kinetic energy per unit mass. This gives for the orbital velocity at any point in the galactic disk for $r > r_s$,

$$v(r) = v_0 \pm \sqrt{r + r_s \ln(r - r_s) + \frac{2GM}{r}}. \qquad (4.4)$$

For the zero of velocity v_0, we have $v_0 = -\int_{v_1(0)}^{v_2(r_s)} dv$. Since $v = \sqrt{E_B}$, therefore the contribution from $r = 0$ to $r = r_s$ is $v_0 = -\sqrt{E_B} \,|_0^{r=r_s} = \sqrt{E_B(0)} - \sqrt{E_B(r_s)}$. Now since the magnetic field (4.1) changes sign across the Schwarzschild radius r_s, we have $E_B = 0$ at $r = r_s$. Using (4.3) we have $v_0 = \sqrt{r + r_s \ln(r_s - r)} \,|_{r=0}$. We therefore obtain $v_0 = \sqrt{r_s \ln r_s}$, which determines the magnitude of v_0 in equation (4.4). Notice that the square-root sign in equation (4.4) is decided by co-rotating and counter-rotating motion in the galactic disk relative to the rotation of the central compact object.

A set of typical galactic curves is shown in Figure

4.1 for the disk region $r > r_s$, where the data points lie, within observational error, on the velocity curves derived from equation (4.4). In plotting the rotation curves for NGC 1097, NGC 1566, and $M31$ galaxies, we have used $E_B = \epsilon_B/B_0^2$. Also the potential energy is scaled per kpc in the galactic disk. In the case of the Triangulum galaxy ($M33$) there is no central bulge, and so the Newtonian contribution to the velocity via GM is negligibly small. The galactic dynamics is thus dominated by the magnetic vorticity term with $B_0 = 0.18\mu G$. This produces the characteristic steepness in the rotation curve of $M33$ galaxy, since $v(r)$ essentially corresponds to $B_0\sqrt{r}$ in this case.

The sign of velocity in equation (4.4) can change for the negative square-root according to if v_0 is greater than or less than the square-root term, this can divide the galactic plane into two distinct regions rotating opposite to each other. Such phenomena has been observed in NGC 4826 and a number of other galaxies. As shown in Figure 4.2, the data points below radius of one kpc lying on the inner disk correspond to velocity in opposite direction to points lying on the scale larger than a kpc.

In galaxies with well formed spiral structure, such

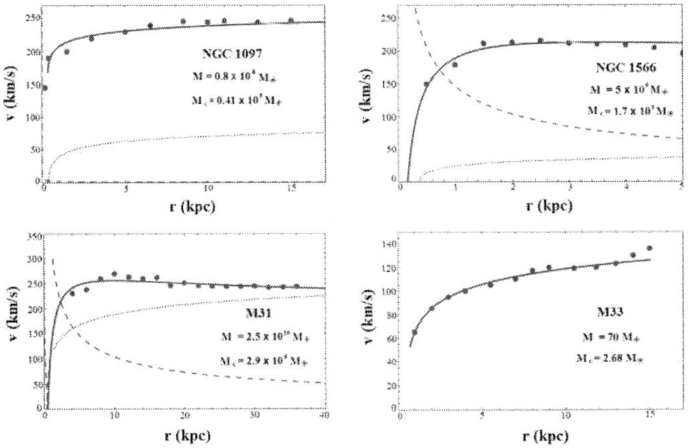

Figure 4.1: Rotation curves (solid plots) based on equation (4.4) for typical galaxies NGC 1097 (data points from Ondrechen, Van Der Hulst & Hummel 1989), NGC 1566 (Bottema, 1992), $M31$ (Carignan, et. al. 2006), and $M33$ (Corbelli, 2000). The dashed plots represent the Keplerian velocity, where as the dotted plots correspond to the magnetic vorticity contribution $\sqrt{r + r_s \ln(r - r_s)}$.

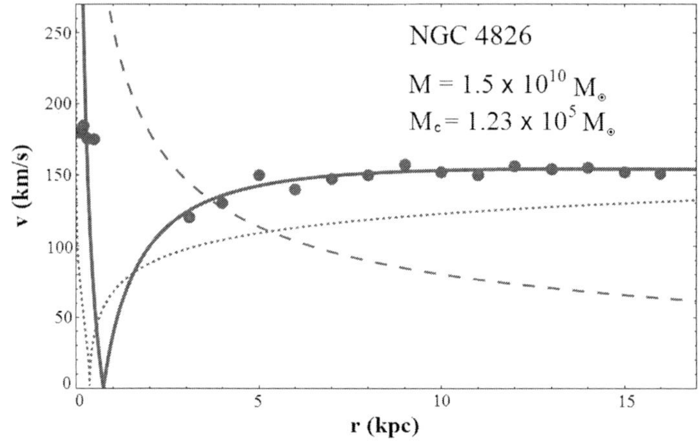

Figure 4.2: Rotation curve for *NGC* 4826 derived from equation (4.4). The plot is shown in the absolute velocity measure, where points on lhs. of the velocity minimum have negative velocity hence represent opposite rotation relative to the points on rhs. of $v = 0$. This corresponds to the counter-rotating disks phenomenon observed in galaxies like *NGC* 4826. Data points are from De Blok, et. al. 2008.

as barred galaxies, mass distribution differs from flat disk galaxies, and is not uniform over large radii. Thus for the Milky Way galaxy (Figure 4.3) we employ the mass distribution given approximately by $M\cos(0.86r)$ in the Newtonian potential. In the context of the present model, we point out that the barred structure of the galactic nucleus corresponds to the points of extrema of the magnetic field (4.1) that differ by an angle π in the galactic plane. Since spacetime dragging is largest at these regions, matter accumulation by accretion is larger at the poles than elsewhere around the central source mass. Once formed, these regions extend in space by the coupled effect of spacetime vorticity and the magnetic energy density build-up around the galactic center by charge accretion.

4.2 The Tully-Fisher Relation

The Tully-Fisher formula for mass-velocity relation can be deduced directly from velocity distribution formula (4.4). In the Tully-Fisher equation maximum luminosity corresponds to maximum rotational speed in the galactic disk. Therefore, by the extremum condition

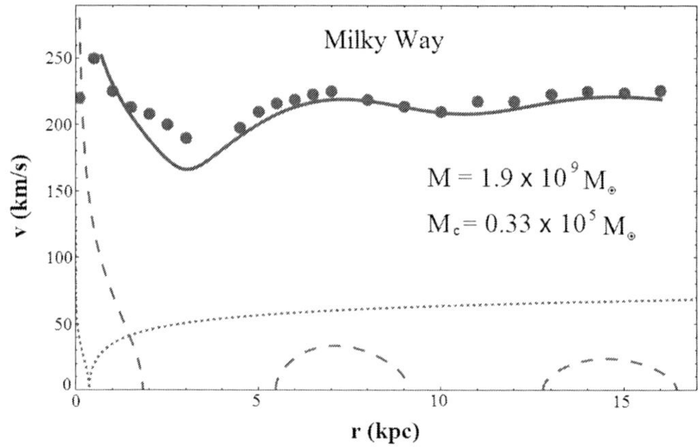

Figure 4.3: Rotation curve (solid plot) for the Milky Way galaxy, with data points from Clemens 1985. The dash and dot plots represent the Keplerian and magnetic vorticity contributions, respectively.

$dv/dr\mid_{v=v_{\max}} = 0$, we have from equation (4.4),

$$r = \frac{2GM}{r} - \frac{2GM}{r^2} r_s, \text{ for } v = v_{\max}. \tag{4.5}$$

Since $r^2 \gg r_s$, for the galactic disk, the second term in equation (4.5) is negligible. Putting therefore in equation (4.4) $r = \sqrt{2GM}$ we find

$$(v_{\max} - v_0)^2 = 2\sqrt{2GM} + r_s \ln(\sqrt{2GM} - r_s). \tag{4.6}$$

By $\sqrt{2GM} > r_s$, the second term on rhs. of equation (4.6) is also negligible compared to the first, we then obtain the Tully-Fisher relation $M \sim v_{\max}^4$. Putting back the magnetic energy density we see that $v_{\max}^4 \sim 8B_0^2 GM$, which shows that for the usual galactic parameters $B_0 \approx 10\mu G$ and total galactic mass of approximately $10^{10} M_{\odot}$, $v_{\max}$ is of the order of $10^2 km/s$, which agrees with the observed order of magnitude of galactic rotational velocities. We notice that when the Schwarzschild radius $r_s = 2GM_c/c^2$, where M_c is the mass of the compact source in galactic center, has magnitude comparable to $\sqrt{2GM}$, deviations from the Tully-Fisher law arise due to the second term on rhs. of equation (4.6).

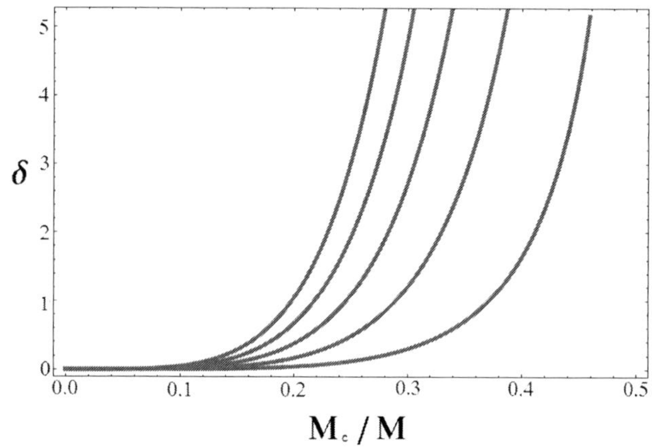

Figure 4.4: Deviations from the Tully-Fisher relation $\delta = \left|(v_{\max} - v_0)^2 - 2\sqrt{2GM}\right|^2$ as function of the central compact mass M_c for different values of the scaled galactic magnetic field. Magnitude of the field B_0 decreases by a unit μG from left to right from $5\mu G$ to $1\mu G$ for the plots. Here we choose the total galactic mass $M = 1$, and in gravitational units $c = 1 = G$.

In Figure 4.4 we give a plot based on equation (4.6) for deviation δ from the Tully-Fisher fourth power law. We see that $\delta \sim B_0^2$ and the predicted deviations from the Tully-Fisher relation become larger for larger magnetic field for a given compact mass, since it increases the maximum velocity in the galactic disk. Also, the deviation δ increases if the central compact source is more massive. But since a larger proportion of the total galactic mass M is then contained in the central compact source than in the luminous matter in the galactic disk, maximum velocity decreases due to the logarithmic term in equation (4.6) which depends on the difference of total mass M and the mass M_c. Since the deviation δ corresponds to $v_{\rm max}^4$ which in turn is proportional to maximum luminosity, this implies that even if the central compact source is more massive, deviation from Tully-Fisher relation for luminosity will still be small provided the galactic magnetic field is sufficiently high.

4.3 Mass Estimate

In equation (4.6) we see that the contribution to the total galactic mass $\sim v_{\max}^4$, due to the magnetic vorticity is $8GM$ for the dominant term, whereas the second term is negligibly small. The Keplerian velocity alone however contributes only $2GM$ to $v_{\max}^4$ for the same radius $r = \sqrt{2GM}$. Given that the total galactic mass inferred from Keplerian velocity distribution is M, the rotational velocity (4.4) thus implies an additional mass of $6M$. This corresponds to the missing galactic mass inferred from rotational velocity curves.

4.4 Galactic Age

For the Milky Way galaxy the total angular momentum, inferred from dynamics of double galaxies, is approximately $10^{67} kg \cdot m^2/s$ (Karachentsev 1987). The time it takes to extend this momentum up to a galactic radius of $r \simeq 10^{18} m$ by the coupled field dragging is given by $t = 2\pi/\omega = 2\pi c^2 r^3/GJ \approx 10^{17} s$ which agrees well with galactic age inferred from observation (Bond, et. al. 2013).

The galactic size can also be determined by knowing the angular speed ω induced by the magnetic field dragging. The average speed of the solar system in the galactic disk is about $550km/s$, this gives for the radius of the Milky Way galaxy, the angular speed around the galactic disk $\omega \approx 10^{-16}/s$, therefore $r = (GJ/c^2\omega)^{1/3} \approx 10^{18}m$, which is of the observed order of magnitude of galactic radius.

Independently of the above formula, one can derive the galactic radius directly from the galactic magnetic field and the speed of stars in the galactic disk. Using from above, $r \approx (v - v_0)^2/B_0^2$ for galactic magnetic field $B_0 \approx 10\mu G$ and the speed induced $v - v_0 \approx 10m/s$ in the galactic disk , gives $r \approx 10^{20}m$ which is again of the observed order of magnitude of the galactic radius (Reid 1993). In general the induced speed v in the disk is higher than or almost equal to the speed v_0 at the galactic center due to the compact source. The difference in speeds is due to the effectiveness of the magnetic field induced dragging, the more such a dragging is effective at larger distances the greater the difference between the two speed will be, and as a result, the larger will be the radius of the rotating galactic disk.

Chapter 5

Conclusions

Rotation curves of galaxies show a marked difference from the Keplerian velocity $\sqrt{2GM/r}$, for galactic mass M and radial distance r, which becomes increasingly significant over the galactic radius. This rotation curve discrepancy is now well established through a large number of astrophysical observations (Sofue & Rubin 2001, De Blok et. al. 2008, Blanton & Moustakas 2009, Naab & Ostriker 2017). A main aspect of galactic dynamics is the diversity exhibited by these rotation curves. Most of the spiral galaxies, like *NGC* 1566, have an almost flat rotation curve at large distances within the galactic disk, where as galaxies like NGC 1097 shows a marked rise in the rotational velocity. Satellite galaxies, like $M31$, even show a sharp steep in rotational speed over the galactic disk.

Despite the observed diversity of rotational patterns of different galaxies, certain features also emerge as similar in general galactic dynamics. For instance, apart from an increase over the Keplerian velocity at large distances, an almost direct proportionality exists between the galactic mass and the fourth power of the maximum rotational velocity. This Tully-Fisher relation (Tully & Fisher 1977, McGaugh et. al. 2000, Verheijen 2001, Kannappan, Fabricant & Franx 2002, McGaugh 2005) is a general scaling relation with no significant dependence on other galactic properties like galaxy size or surface brightness, and has little intrinsic scatter (Courteau & Rix 1999, McGaugh 2005, Sorce & Guo 2016, McGaugh, Lelli & Schombert 2016, Lelli, et. al. 2017).

In this work, we have derived a model of galactic dynamics based on first principles, namely Maxwell equations in curved spacetime. We have deduced main features of galactic motion from directly observable galactic parameters, namely the galactic mass, radius, and the magnetic field. In this context the magnetic field generated in galaxies, mainly by accretion around galactic nuclei, provides the feedback mechanism for galactic rotation. Analysis of general relativistic Maxwell equations shows that toroidal magnetic field thus generated enhances spacetime vorticity. Such a coupling occurs when the central mass has non-zero angular momentum and an intrinsic magnetic field is present in the galactic center. Spacetime dragging is thereby extended to much larger distances under the magnetic pressure. We show that this enhanced spacetime dragging corresponds directly to the velocity increase observed in the diverse behavior of galactic rotations.

The model presented here depends on the following main features of galactic structure:

(1) A large proportion of galactic mass is contained in the region in and around the galactic center. In most spiral galaxies this region is fully formed as galactic bulge, and despite very different morphologies, masses, sizes, and gas fractions; matter in these galaxies can be generally divided into a disk and a central bulge or a nuclear mass.

(2) Although most disk galaxies have a central bulge, pure disk galaxies contain a compact source at the

galactic center. In such cases galactic dynamics is dominated by the compact source. In general, the existence of central compact gravitational sources is now well established in almost all galaxies (Heckman & Best 2014, Reines & Volonteri 2015, Terrazas, et. al. 2017, Bogdán, et. al. 2018).

(3) Galaxies have an intrinsic magnetic field, which we identify as toroidal field generated in the central region. In spiral galaxies the galactic magnetic field strength varies from $10 - 15\mu G$ within the galactic arm (Beck 2001, Van Eck, et. al. 2015, Beck 2016). Observationally, the total magnetic field strength can be determined from the intensity of total synchrotron emission, assuming energy balance (equipartition) between magnetic fields and cosmic rays. The typical average equipartition strength for spiral galaxies is about 10 μG. In general the galactic magnetic field has been observed to be of order of μG which extends across the galactic plane (Pakmor, Marinacci, & Springel 2014, Van Loo, Tan & Falle 2015, Tabatabaei, et. al. 2016).

Also, in the linear approximation to Einstein field equations there is the gravitomagnetic field which couples to the magnetic field in particle dynamics around rotating magnetized objects (Mirza 2007, also see Mirza

2019, 2020 for further references). The gravitomagnetic force however is too weak to derive orbital motion at large distances. Moreover, its effects decay as $1/r^3$, hence cannot support flat rotation curves. The effect we have discussed here is distinct from Lense-Thirring frame drag, since it results from a coupling of gravitational and electromagnetic fields. Moreover it is generally valid for any axially symmetric matter distribution endowed with a magnetic field structure. In astrophysical systems accretion provides an effective means of magnetic field generation around massive objects. Spacetime around a rotating compact object, while remaining locally isomorphic to the Kerr spacetime, thus modifies the dynamics of matter by the mechanism discussed above. The formation of central region of sufficient mass density and magnetic field strength is essential for observed rotational dynamics of galaxies. In ellipticals, disk-like region formation in the galactic centers due to rotation have also been observed (Rix 1990, Naab & Burkert 2001), indicating that the toroidal field coupled to the spacetime vorticity is also important in the evolution of spiral and elliptical galaxies. The above model leads to the diverse rotation curves of dwarf galaxies (Oman, et al.

2015), where the magnetic field generated by spacetime vorticity is comparable to that in disk galaxies. Also, the counter-rotation of inner discs observed in some galaxies , as in *NGC* 4826 (Fig. 4.2), cannot be accounted for by non-Newtonian accretion and various dark matter scenarios.

Bibliography

[1] Beck, R. 2001, *SSRv*, 99, 243

[2] Beck, R. 2016, *A&ARv*, 24, 4

[3] Blanton, M. R., & Moustakas, J. 2009, *ARvA&A*, 47, 159

[4] Bogdán, Á., Lovisari, L., Volonteri, M., & Dubois, Y. 2018, *ApJ*, 852, 131

[5] Bond, H. E., Nelan, E. P., VandenBerg, D. A., Schaefer, G. H., & Harmer, D. 2013, *ApJL*, 765, L12

[6] Bottema, R. 1992. *A&A*, 257, 69

[7] Carignan, C., Chemin, L., Huchtmeier, W. K., & Lockman, F. J. 2006, *ApJL*, 641, L109

[8] Ciufolini, I., 1989, *IJMPA*, 4, 3083

[9] Corbelli, E., & Salucci, P. 2000, *MNRAS*, 311, 441

[10] Clemens, D. P. 1985, *ApJ*, 295, 422

[11] Courteau, S., & Rix, H. 1999, *ApJ*, 513, 561

[12] De Blok, W. J. G., Walter, F., Brinks, E., Trachternach, C., Oh, S. H., & Kennicutt Jr, R. C. 2008, *AJ*, 136, 2648

[13] Eddington, A.S. 1919, *The Observatory*, 42, 119

[14] Einstein, A., 1912, *Viertelj. Schrift. Ger. Medizin*, 44, 37

[15] Einstein, A., Grossmann, M., 1913, *Z. Math. Physik*, 62, 225

[16] Einstein, A. 1952, *Principle of Relativity*, Dover Publications

[17] Heckman, T. M., & Best, P. N. 2014, ARvA&A, 52, 589

[18] Hubble, E. 1926, *Contributions from the Mount Wilson Observatory/Carnegie Institution of Washington*, 324, 1

[19] Iorio, L. 2010, *New Astron.* 15, 554

[20] Kannappan, S. J., Fabricant, D. G., & Franx, M. 2002, *ApJ*, 123, 2358

[21] Karachentsev, I. D. 1987, *Double galaxies*, Izdatel Nauka, Moscow

[22] Kerr, R.P. 1963, *Phys. Rev. Lett.* 11, 237

[23] Lämmerzahl, C., Everitt, C.F. & Hehl, F.W. eds., 2001, *Gyros, clocks, interferometers...: testing relativistic gravity in space (Vol. 562)*, Springer Science & Business Media

[24] Lelli, F., McGaugh, S. S., Schombert, J. M., & Pawlowski, M. S. 2017, *ApJ*, 836, 152

[25] Lense, J., and Thirring, H. 1918, *Phys. Z.*, 19, 156

[26] Lichnerowicz, A., 1967, *Relativistic Hydrodynamics and Magnetohydrodynamics,* Benjamin Press, New York

[27] Mach, E., 1893. *The Science of Mechanics: A Critical and Historical Exposition of its Principles,* Open Court Publishing Company.

[28] McGaugh, S.S., Schombert, J. M., Bothun, G. D., & de Blok, W. J. G. 2000, *ApJ.* 533, L99

[29] McGaugh, S. S. 2005, *ApJ*, 632, 859

[30] McGaugh, S. S. 2005, *PRL.* 95, 171302

[31] McGaugh, S. S., Lelli, F., & Schombert, J. M. 2016, *PhRvL*, 117, 201101

[32] Mirza, B.M. 2007, *IJMPD*, 16, 1705

[33] Mirza, B.M. 2017, *ApJ*, 847, 73

[34] Mirza, B.M. 2019, *MNRAS*, 489, 3232

[35] Mirza, B.M. 2020. *ApSS*, 365, 1

[36] Naab, T., & Burkert, A. 2001, *ApJL*, 555, L91

[37] Naab, T., & Ostriker, J. P. 2017, *ARvA&A*, 55, 59

[38] Oman, K. A., Navarro, J. F., Fattahi, A., Frenk, C. S., et al. 2015, *MNRAS*, 452, 3650

[39] Ondrechen, M. P., Van Der Hulst, J. M., & Hummel, E. 1989, *ApJ*, 342, 39

[40] Oort, J. H. 1932, *Bull. Astron. Inst. Neth.* 6, 249

[41] Oron, A., 2002, *Phys. Rev. D*, 66, 023006

[42] Pakmor, R., Marinacci, F., & Springel, V. 2014, *ApJL*, 783, L20

[43] Pétri, G., 2004, *Class. Quantum Grav.* 21, 3433

[44] Reines, A. E., & Volonteri, M. 2015, *ApJ*, 813, 82

[45] Reid, M. J. 1993, ARA&A, 31, 345

[46] Rindler, W., 2006, *Relativity: special, general, and cosmological*, Oxford University Press, 2nd edition

[47] Rix, H. W., & White, S. D. 1990, *ApJ*, 362, 52

[48] Rubin, V.C., 1983, *Science*, 220, 1339

[49] Ruggiero, M.L. & Tartaglia, A., 2002, *arXiv preprint gr-qc/0207065*

[50] Sofue, Y., & Rubin, V. 2001, *ARA&A*, 39, 137

[51] Sorce, J. G., & Guo, Q. 2016, *MNRAS*, 458, 2667

[52] Tabatabaei, F. S., Martinsson, T. P. K., Knapen, J. H., Beckman, J. E., Koribalski, B., & Elmegreen, B. G. 2016, *ApJL*, 818, L10

[53] Terrazas, B. A., Bell, E. F., Woo, J., & Henriques, B. M. 2017, *ApJ*, 844, 170

[54] Tully, R. B., & Fisher, J. R. 1977, *A&A*. 54, 661

[55] Van Eck, C. L., Brown, J. C., Shukurov, A., & Fletcher, A. 2015, *ApJ*, 799, 35

[56] Van Loo, S., Tan, J. C., & Falle, S. A. 2015, *ApJL*, 800, L11

[57] Verheijen, M. A. W., 2001, *ApJ*, 563, 694

[58] Weber, F., & Glendenning, N. K. 1992, *ApJ*, 390, 541

[59] Will C. M., 2001, *Living Reviews in Relativity*, 4, 4

[60] Zwicky, F. 1937, *ApJ*, 86, 217